THE MOTORCYCLE

1963
rton Atlas 750cc
1965
BMW R50/2 500cc

THE MOTORCYCLE

120 YEARS OF MOTORIZED MASTERPIECES FROM THE HAAS MOTO MUSEUM

weldonowen

Indian

CONTENTS

HAAS
MOTO MUSEUM
& SCULPTURE GALLERY
MUSEUM PARKING
MUSEUM

FOREWORD

"The hallmark of a great partnership is when each of you thinks you're the lucky one."
—Bobby Haas

The first day I met Bobby, we sat around a large conference table with a few other people, and he made a bold statement: "We are going to build the finest moto museum that exists, pound for pound. And anyone that is not on board with that mission can leave the room." Needless to say, none of us moved a muscle.

When I walked into the meeting that day, the first motorcycle I saw was The Rocket, an exquisite custom bike designed by Bobby and built by Sparky Williamson and the crew at Strokers Dallas (see pages 154–55). It was the most exquisite color green and the most captivating machine I've ever seen. Bobby's passion for motorcycles was infectious, and I was immediately intrigued and hooked. I had no background in motorcycles, nor did I know how much I would grow to embrace them so deeply in my life, but I knew I wanted to help Bobby fulfill his vision. Our partnership began that day.

Bobby and I first shared the Haas motorcycle collection with the public at a 5,000-square-foot facility called the Haas Motorcycle Gallery at Dragon. At the time it was only around fifty motorcycles, with the focus primarily on the vintage cycles. Once the Dragon was filled, Bobby did not want to stop collecting; his hunger to find and curate a profound and world-class collection was endless. Traveling the world, meeting wonderful people along the way, we came to the realization that the collection was as much about the stories of these bikes and the relationships people had to them as it was about the bikes themselves.

We quickly outgrew the Dragon, so we found a 25,000-square-foot space, which is the current home of the Haas Moto Museum. Then we had to figure out how to fill it. We needn't have worried—the museum began to tell us what it wanted to be, and we listened with intent. For the most part, Bobby's collection had been centered on vintage and historically important motorcycles, but once we opened the museum, we began to meet custom builders from all over the world. Bobby realized that not only were custom-built machines needed in the collection, but he also wanted to support these builders' artistry and become more involved himself in the custom motorcycle world. And so, the Custom Shop was born. We had no idea that not only would the partnerships with these builders be so fulfilling, but also that the custom collection would prove to be the pièce de résistance of the museum. We traveled some more, made new connections, and curated the most outstanding collection of custom motorcycles in the world. Our passion for the motorcycles and the people behind them runs deep in our veins.

Previous pages: ***The Rocket*** *by Bobby Haas and Strokers Dallas* (*see pp. 154–55*)

This book pays tribute to a museum like no other, a diverse collection thoughtfully curated by a man like no other. Although Bobby was taken from us too soon, he is very much a part of this book. The words you will read about each of the motorcycles were written by him, and stand proudly next to each bike in the museum. As you peruse the pages, I hope that you can feel Bobby's spirit guiding you through the museum. It is strong and it is loud.

Bobby and I were partners in every sense of the word, in love and life. I believe that anybody who had the privilege of knowing Bobby Haas is a lucky individual, and even those who never had the honor to meet him are and will continue to be inspired by his passions. There will never be another Bobby Haas, for that I am certain, because he is one of a kind.

—Stacey Mayfield, *Co-founder and Museum Director, Haas Moto Museum*

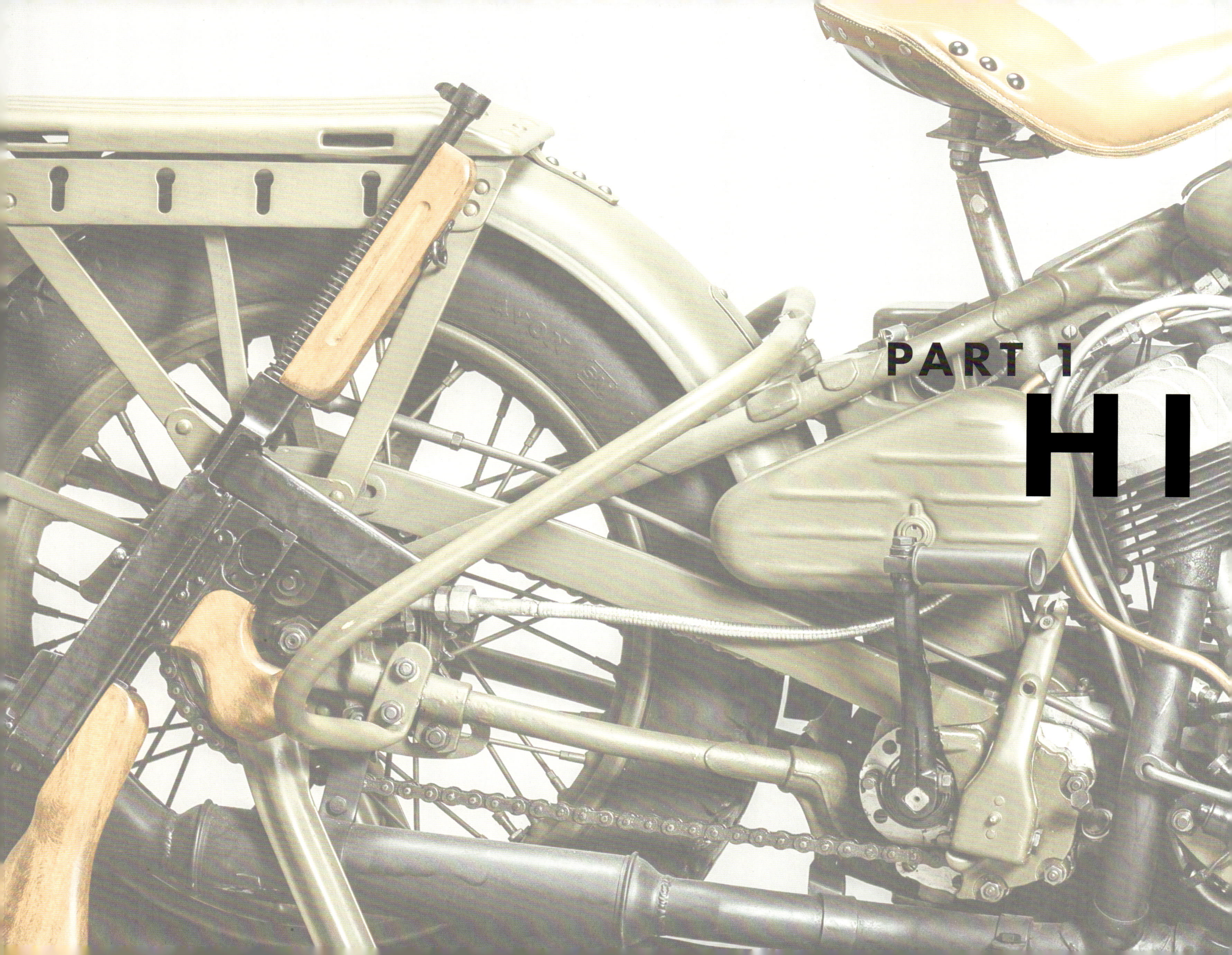

PART 1
HI

STORY HALL

COLLECTING HISTORY'S STORIES

I've often been asked how I curated this museum ... why I chose certain bikes or certain pieces of artwork over others to occupy a place in the museum. After all, I didn't have any experience as the curator of a museum and barely any experience with motorcycles ... but somehow, the museum and the entrance into History Hall have managed to come together in a way that appeals to a wide variety of guests, from moto experts to people who barely know how a motorcycle works.

When I first started collecting, I relied heavily on the fact that a certain bike would fill a niche that needed to be filled in the lineup. But after a while, I realized that if I was going to carve my own niche in this industry, I needed to rely on an approach that I knew a little something about...and that something was my decade-long career as a photographer for *National Geographic.*

During my time working with *Nat Geo*, I was constantly reminded to bring back images that tell a story ... or "bring back images unlike anything seen before." I was told that if I was going to create a magazine like no other, a book like no other, an exhibit like no other, I needed something more than beautiful images—I needed images that told a story ... images that forced readers to dig deep and wonder what the story was behind each image.

And that's exactly how I designed History Hall—with a passion to display motorcycles and stories rarely seen or heard before.
—Bobby Haas

History Hall offers a leisurely stroll through the entirety of motorcycle history, arranged in chronological order from late 19th-century bicycle frames with attached gas tanks to the throbbing power stations of ultra-modern road warriors. The cavernous contours of History Hall afford each cycle ample room to breathe aboard its custom-designed platform. Regardless of whether you are a newcomer to this world or a moto aficionado, the diversity of more than ninety motorcycles in History Hall leaves you breathless, convinced that no other place in the pantheon of motorcycle museums may boast such a collection.

1899 PEUGEOT TRICYCLE 2-$^{1}/_{4}$HP

FRANCE

Powered by the famous De Dion Bouton single-cylinder engine, this late 19th-century Peugeot was painstakingly restored in Holland by Gilbert Warning. This testament to the pioneer days of motorcycling is exceedingly rare and genuine, retaining many of the original components, most notably its De Dion engine and Peugeot two-speed gearbox.

Clockwise, from top left:

c.1901 L'UNIVERSEL 175CC

FRANCE

The Grande Dame of the Haas Collection and perhaps the last survivor of this obscure brand.

1902 CENTAUR 211CC

UNITED KINGDOM

Featuring the distinctively shaped Minerva clip-on engine, this may very well be the oldest Centaur still in existence.

c.1903 CLÉMENT MODEL D 192CC

FRANCE

Cylindrical petrol tank positioned behind the saddle, with batteries in a leather case strapped to the frame.

Clockwise, from top left:

1903 MINERVA CYCLE AND MILLS AND FULFORD FORE-CAR 239CC

BELGIUM

Meticulous restoration brings to life an early effort to place the passenger directly in front of the cycle; the fore-car was detachable, allowing the front wheel to be reattached in a more conventional configuration.

1904 REX 373CC

UNITED KINGDOM

Anointing itself as "The King of British Motors," Rex was a design leader, with features like the distinctive L-shaped gas tank and a matching battery and toolbox directly beneath the saddle.

1910 MOTOSACOCHE MT 225CC

SWITZERLAND

Motosacoche translates roughly as "engine in a saddlebag," a fitting moniker for this early cycle whose motorized components fit neatly within the bicycle frame's a center triangle.

Opposite: 1904 ***Rex***

1904

Clockwise, from top left:

1911 ZENITH 500CC

UNITED KINGDOM

Famous for its Gradua gear shift with sliding rear wheel to adjust the belt length to different settings.

1911 PEUGEOT MOTO LEGERE 350CC

FRANCE

Elegant V-twin with removable headlamp that doubled as a lantern.

1912 INDIAN HEDSTROM SINGLE 497CC

UNITED STATES

The rearward-sloping single-cylinder engine designed by Oscar Hedstrom is cradled inside a loop frame previously adopted as the frame of choice by archrival Harley-Davidson. Though crude by modern standards, the hand-operated shifter was considered a major advance in its day. The gold script lettering of the name "Indian" on the fuel tank and its brilliant red coat became iconic marks of this distinctly American brand.

Clockwise, from top left:

1914 WALL AUTOWHEEL 118CC

UNITED KINGDOM

Motorized third wheel propels this woman's bicycle frame.

1915 BRADBURY 6HP V-TWIN 749CC

UNITED KINGDOM

Stunning survivor of a rare marque destined to cease production in 1925.

1916 INDIAN POWERPLUS 1000CC

UNITED STATES

The passage of more than a century has not dulled the luster of this precocious 1000cc V-twin with multiple sets of leaf springs and dazzling accent features.

Overleaf: 1912 ***Wall Autowheel*** *(left), 1918* ***FN 285*** *(right)*

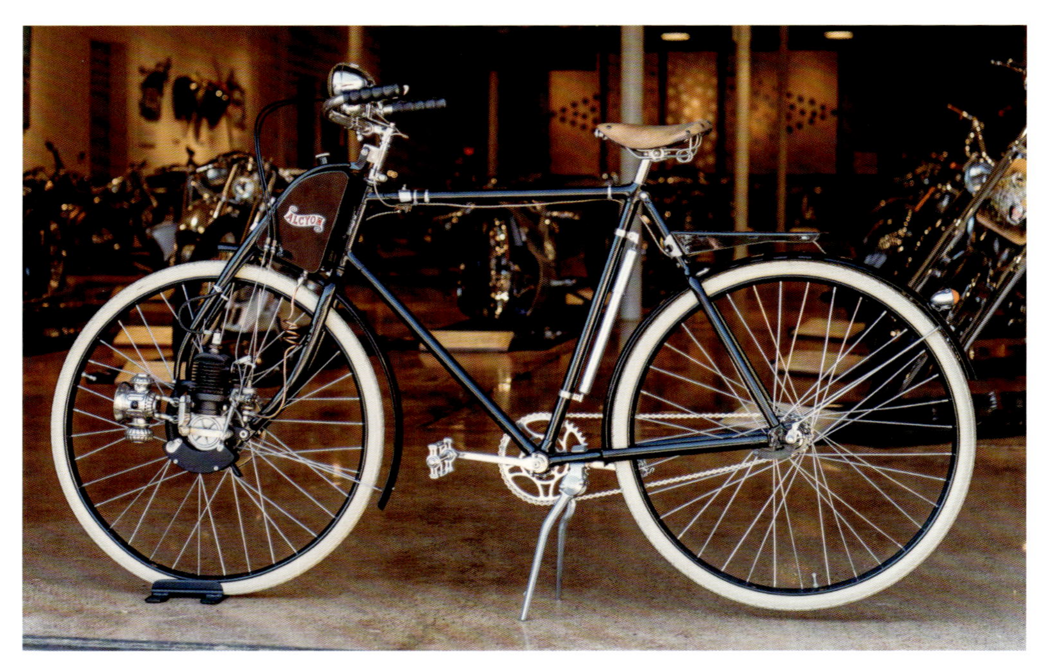

Clockwise, from top left:

1917 CLEVELAND A2 LIGHTWEIGHT 221CC

UNITED STATES

Built over a century ago, this two-stroke single features an unusual configuration with the crankshaft in-line with the frame and a cylindrical-shaped gas tank. Beautifully restored "from the ground up," all parts on this cycle are believed to be original, save for the wheel rims and spokes.

c.1918 FN 285 285CC

BELGIUM

Century-old Belgian model delivers power to its rear wheel with distinctive shaft drive.

1919 ALCYON CYCLO MOTEUR 2HP

FRANCE

One of only two known survivors of the Alcyon 2HP model.

1919

ABC SKOOTAMOTA

125CC

UNITED KINGDOM

Spacious footboard with a single cylinder mounted above the rear wheel; one of the earliest moto scooters.

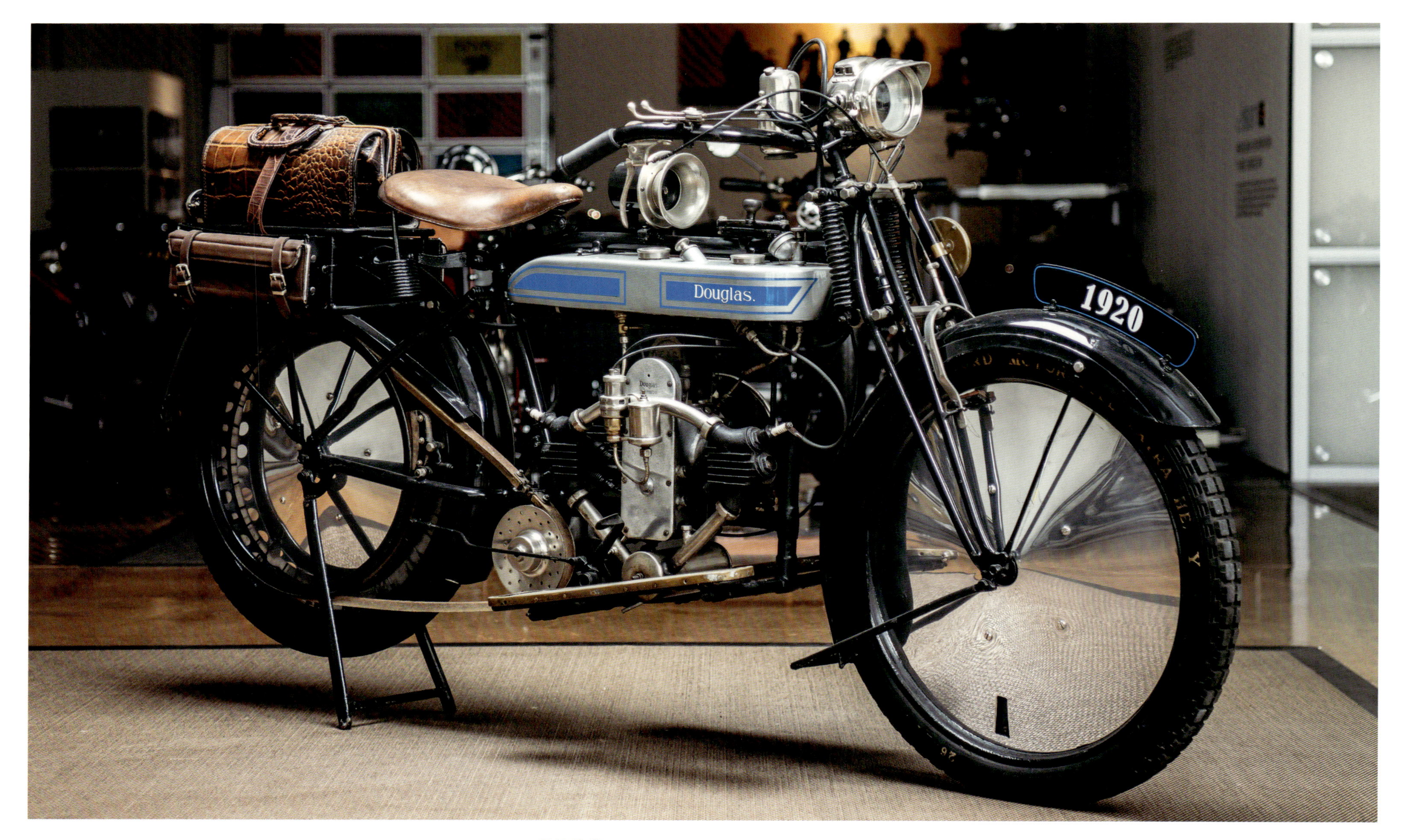

c.1920 DOUGLAS MODEL WD20 2-3/4 HP 350CC

UNITED KINGDOM

Outfitted for a ride in the country and accented with fashionable aluminum disc wheels of its era.

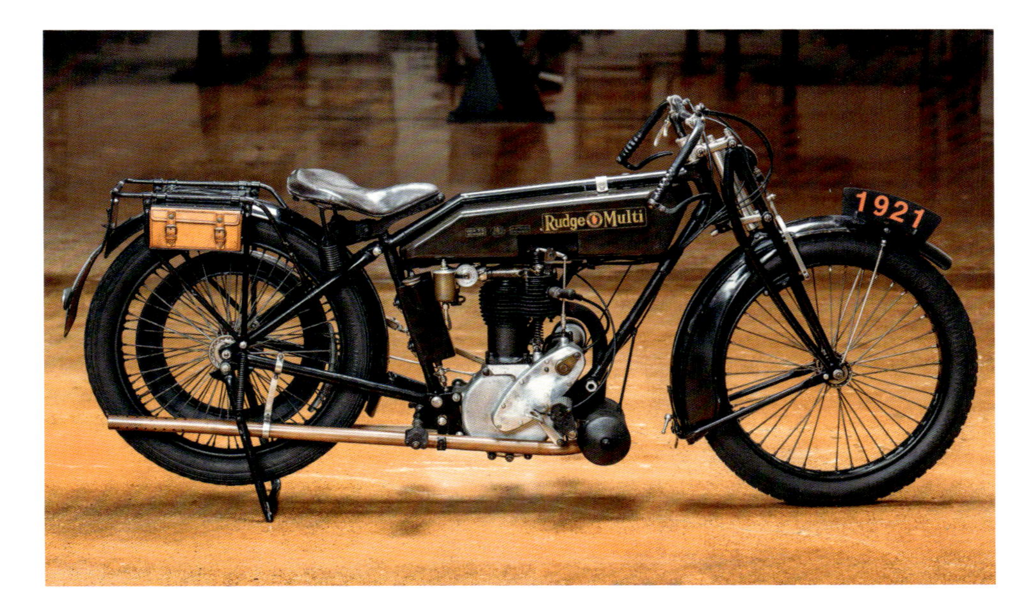

Clockwise, from top left:

1920 EXCELSIOR SERIES 20 BIG TWIN DELUXE 1000CC

UNITED STATES

With its century anniversary in 2020, this exquisitely restored Excelsior Series 20 Deluxe edition features the big twin 1000cc engine, all-chain drive, curved top tube with rounded gas tank, and the brilliant blue color introduced in 1920 to replace Excelsior's traditional olive coat.

1921 RUDGE MULTI TT 500CC

UNITED KINGDOM

With a heritage that includes both racing success and service during World War I, the Multi version features continuously variable transmission with reportedly 64 separate clutch plates.

1922 HARLEY-DAVIDSON MODEL WF 584CC

UNITED STATES

A beautifully restored example of one of the rarest of Harley-Davidson "flat twins," or horizontally opposed twins with the engine set lengthwise in the frame. The 6hp twin was installed in a so-called "keystone" frame, open at the bottom and thus using the engine as a stressed element. Introduced in 1919, production of the W was halted in 1923, leaving only a few survivors of this model almost one century later.

Clockwise, from top left:

1923 HENDERSON DE LUXE 1301CC

UNITED STATES

With its trademark four-cylinder engine, the Henderson was the speed merchant of its time, favored by police departments for being the fastest vehicle on the road.

1924 NUT V-TWIN 700CC

UNITED KINGDOM

Hailing from the small English town of Newcastle upon Tyne (abbreviated NUT), this marque gained fame when Hugh Mason, one of its two founders, survived a vicious crash on a practice run and yet eventually prevailed as winner of the 1913 Isle of Man Junior TT. NUT produced fewer than 400 machines in its twenty-year history, and this classic brown survivor features the company's signature round gas tank fastened with drilled metal straps.

1924 HARLEY-DAVIDSON MODEL JE 1000CC

UNITED STATES

Signature curved notches in the gas tank allow breathing space for the twin cylinders.

Opposite: 1923 ***Henderson De Luxe***

EXELSIOR
MOTOR
MFG
HENDERSON
AND
SUPPLY CO
CHICAGO, USA

1925 MAGNAT-DEBON ENTRETUBE SUPER CONFORT 350CC

FRANCE

Stunning restoration of a nearly century-old sport comfort model of Magnat-Debon, which by 1925 was a "sous-marque" of the Terrot cycle and motorcycle empire.

Clockwise, from top left:

1926 PEUGEOT P104 350CC

FRANCE

Unusual leg guards accent this Peugeot Model P104 with "entretube" between the frame tubes petrol tank; production was limited to only 2,000 units, and survivors are few and far between.

c.1926 NER-A-CAR 255CC

UNITED STATES

Radical design reminiscent of the early autos of its era with a low-slung frame, hub-center steering, and long wheelbase.

1928 EXCELSIOR SUPER-X 750CC

UNITED STATES

Cloaked in its classic military green, Excelsior faced fierce competition from American rivals Harley-Davidson and Indian; the brand eventually succumbed to the Great Depression in 1931.

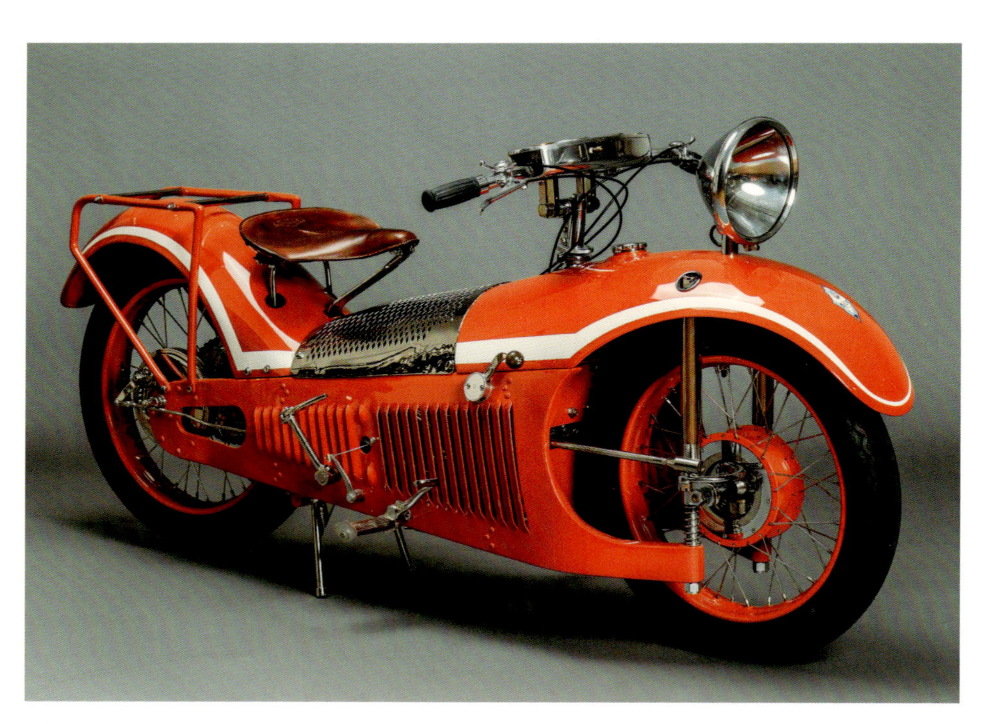

Clockwise, from top left:

1928 GILLET HERSTAL 350 SPORT 350CC

BELGIUM

Along with its Belgian brethren FN and Saroléa, Gillet Herstal produced premier racers during the pre-WWII era, including this 350cc Sport model.

1929 NSU 201T SPORT 199CC

GERMANY

With a heritage that stretches back to the 19th-century manufacture of knitting machines, NSU eventually found its way onto the upper echelons of racing prominence.

1929 MAJESTIC 500CC

FRANCE

One of the few remaining survivors of this magnificent brand, painstakingly restored to its original Art Deco glory by its previous owner.

Opposite: 1929 ***Majestic***

MAJESTIC

Clockwise, from top left:

1929 NEW-MOTORCYCLE 500CC

FRANCE

Side-by-side with the Majestic, another radical George Roy-designed creation with a pressed-steel chassis.

1929 MOTOSACOCHE 250 CM3 250CC

SWITZERLAND

The brand dates back to the 1900 production of an auxiliary engine for a bicycle; the name *Motosacoche* means "engine."

1929 NEANDER 1000CC

GERMANY

An exceedingly rare and innovative vehicle conceived by Ernst Neumann-Neander, this motorcycle offers a radical Art Deco design with a curved seat and egg-shaped fuel tank mounted above a pressed-steel frame that was cadmium plated. This Neander is powered by a 1000cc JAP V-twin side-valve engine with a hand-shifted three speed gearbox and Bosch magneto.

Opposite: 1929 ***Motosacoche 250 CM3***

MOTOSACOCHE
GENÈVE - SUISSE

GH 3858

Clockwise, from top left:

c.1929 **ASCOT-PULLIN** 499CC

UNITED KINGDOM

One of the few survivors of the Art Deco masterpiece designed by Isle of Man TT champion Cyril Pullin, this cycle was expertly restored by Colin Light for his brother Derek Light, a renowned British collector. The cycle boasts a bevy of innovative features, including a pressed-steel chassis, horizontally mounted single-cylinder engine, and exquisite dashboard gauges, to name just a few.

1930 **MOTOSACOCHE 413 TOURISME** 500CC

SWITZERLAND

Its sleek aesthetics envelop the proprietary MAG Motosacoche Acacias Genève engine, long sought after by other European manufacturers.

1930 **CONDOR 322 POPULAIRE** 500CC

SWITZERLAND

Though not nearly as "populaire" as its name would imply, the Condor line carried the unmistakable mark of refined Swiss styling.

Opposite: 1929 ***Ascot-Pullin***

Clockwise, from top left:

1930 HARLEY-DAVIDSON MODEL V 1200CC

UNITED STATES

The "flathead" V-twin, considerably heavier than its predecessors, but with the flexibility of interchangeable wheels and detachable cylinder heads.

1930 TRIUMPH SSK 350 350CC

GERMANY

A product of Triumph-Werke Nürnberg, a German company founded in 1896 by Siegfried Bettmann, who also founded the UK Triumph bicycle and motorcycle line in Coventry, England.

c.1930 NEW MAP JT3 350CC

FRANCE

A sporty French model renowned for its handsome looks and bevy of different engines, in this case a JAP, JA Prestwich single cylinder.

Opposite: 1930 ***New Map JT3***

FINA

c.1931 MGC MODEL N3C 350CC

FRANCE

Revolutionary frame of aluminum alloy castings with the gas and oil tanks integrated into the frame. An MGC was chosen to be featured in the famous Guggenheim "Art of the Motorcycle" exhibit; this cycle is one of only six known Model N3C survivors out of thirty-seven ever produced.

Clockwise, from top left:

1931 DRESCH MONOBLOC 500CC

FRANCE

Popular with the Parisian gendarmes, but a casualty of the 1939 downturn brought on by World War II.

1931 DKW BLOCK 200 192CC

GERMANY

An ancestor company of the modern-day Audi, DKW was a world leader in two-stroke design and one of Europe's largest motorcycle producers prior to the outbreak of World War II.

1931 HUSQVARNA MODEL 50 TV 500CC

SWEDEN

Powered by an OHV JAP engine and outfitted with a Sturmey-Archer gearbox and heavy parallelogram forks, this Husqvarna Model 50 TV was treated to a 334-hour restoration in its native Sweden, turning back the clock to virtually mint condition. Only 408 Model 50s were ever produced, and this Model 50 TV is believed to be the only one in the U.S. at the time of its recent purchase by the Haas Moto Museum.

Overleaf: 1931 ***Husqvarna Model 50 TV*** *(left), 1931* ***Griffon G505*** *(right).*

ENSIGN

Mobiloil

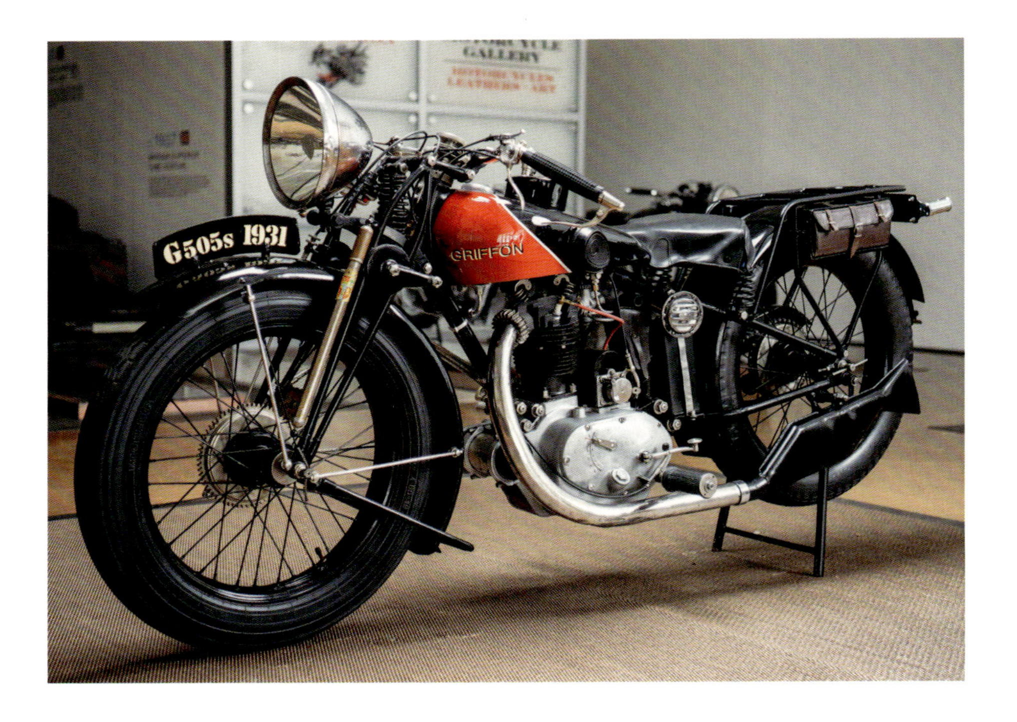

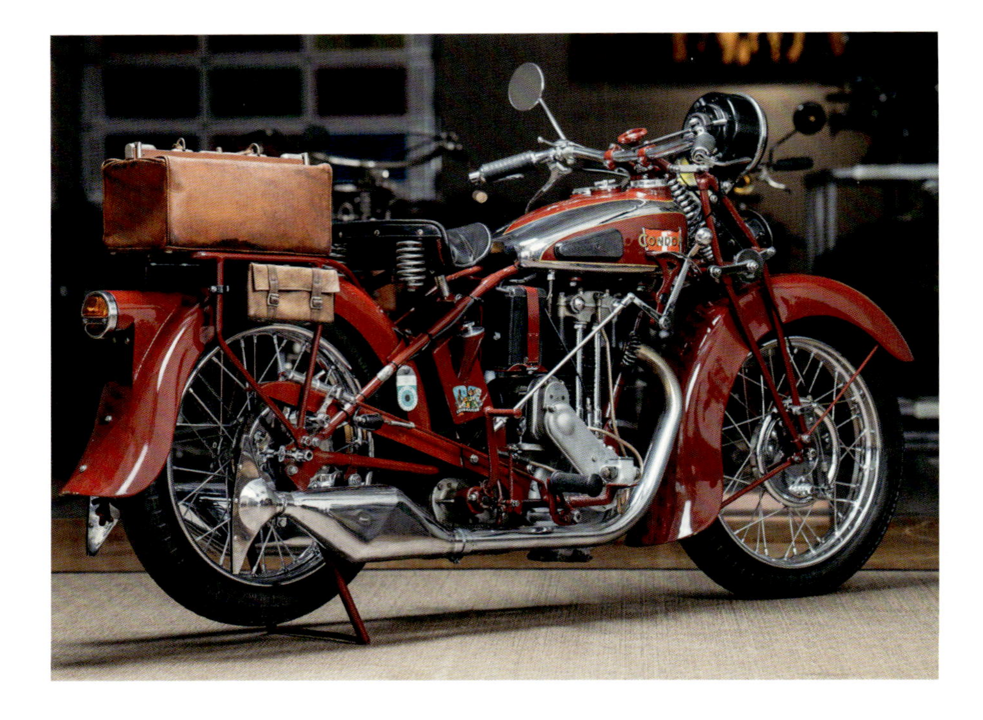

Clockwise, from top left:

1931 **GRIFFON G505** 350CC

FRANCE

Originating in the early 20th century when the Griffon Bicycle Company fitted a Zédel engine on a simple two-wheeled frame, the Griffon motorcycle line soon achieved notoriety as dominant racers before being absorbed by Peugeot in 1927, only four years before this model was produced.

1932 **CONDOR GRAND SPORT** 500CC

SWITZERLAND

Flashy exterior and sporty rear satchel belie Condor's origins as a supplier of drab military cycles.

1932 **PEUGEOT P108** 250CC

FRANCE

Most recognizable for its renowned automobiles, Peugeot's ancestry also includes bicycles, motorized tricycles, and stylish motorcycles such as this.

Opposite: 1932 ***Condor Grand Sport***

CONDOR
PASSO STELVIO
STILFSERJOCH
CONDOR

Clockwise, from top left:

1932 **TERROT VM** 100CC

FRANCE

The Terrot VM was one example of the very economical BMA class "bicyclette avec moteur auxiliare," or bicycle with auxiliary engine, developed during the Great Depression of the 1930s.

1933 **GNOME ET RHÔNE CM1** 350CC

FRANCE

A hardy marque destined to survive both World Wars and the Great Depression, Gnome et Rhône was widely known in the aviation industry for its high-quality aircraft engines; this fine example features dual exhausts emerging from a single upright cylinder.

1933 **PEUGEOT P50T** 100CC

FRANCE

Peugeot's entry in the economical BMA class "bicyclette avec moteur auxiliare," or bicycle with auxiliary engine, was developed during the Great Depression; top speed was 22mph.

Opposite: 1932 ***Terrot VM***

TERROT

NSU
NSU

Clockwise, from top left:

1934 NSU OSL 251 241CC

GERMANY

Exquisitely maintained as it was passed from generation to generation in the same family throughout its entire life, before being acquired by the Haas Moto Museum.

1935 CALTHORPE IVORY 500CC

UNITED KINGDOM

Distinctive ivory finish and upswept exhaust pipes were two of the original options of this member of the Ivory Major line, of which there are only a few survivors.

1936 KOEHLER-ESCOFFIER "MOTO-BALL" 350CC

FRANCE

The chariot of choice in the polo-like sport of Moto-Ball, all the rage in France in the 1930s.

Opposite: 1934 ***NSU OSL 251***

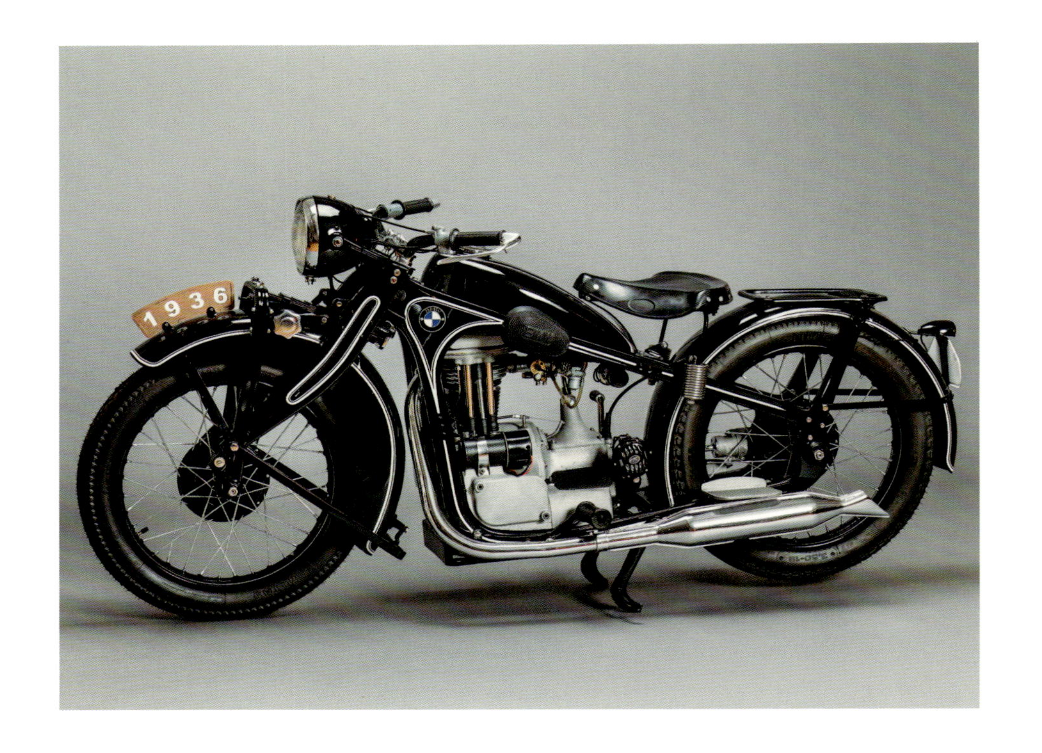

Clockwise, from top left:

1936 BMW R-4 398CC

GERMANY

Its pressed-steel frame and classic sloping lines reveal an upright single cylinder, as opposed to BMW's penchant for producing "boxer" style flat twins.

1937 MERCIER MOTO CHENILLE 350CC

FRANCE

Rejected by the French Army as an impractical track-driven hill climber, this bizarre prototype was destined to charm and mystify audiences for generations.

c.1938 MATCHLESS G80 CLUBMAN 500CC

UNITED KINGDOM

Twin ports and dual high exhausts emerge from the single cylinder powering this rare Matchless Clubman.

Opposite: 1937 ***Mercier Moto Chenille***

JAP

1938 BROUGH SUPERIOR SS80 DE LUXE 982CC

UNITED KINGDOM

The Brough Superior still reigns supreme as the Rolls-Royce of its genre, for its meticulous design, advanced engineering, and raw speed; in its day, the SS80 was often attached to sidecars such as the one exhibited on page 80.

Clockwise, from top left:

1942 HARLEY-DAVIDSON WLA 750CC

UNITED STATES

With many WLAs from World War II available at bargain prices to returning veterans as post-war surplus, the WLA would help spark the biker culture that endures to this day.

c.1942 SOCOVEL ELECTRIC CYCLE

BELGIUM

One of the earliest electric motorcycles ever, this Socovel was one of an edition of only about 400 produced by the Limelette brothers during the Nazi occupation of Belgium during World War II.

1943 NORTON 16H 500CC

UNITED KINGDOM

The workhorse of the British Army during World War II.

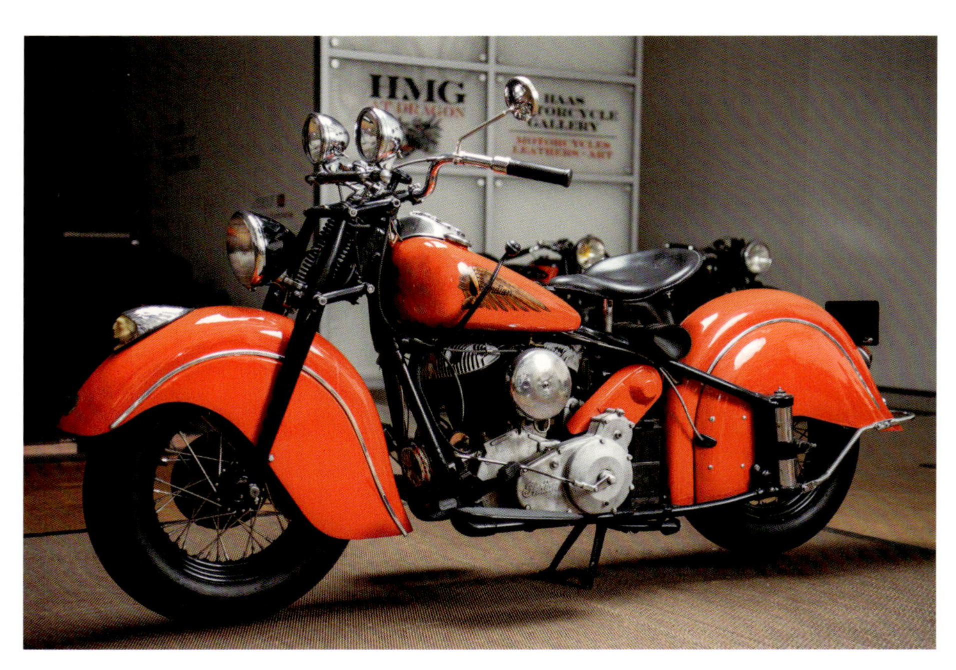

Clockwise, from top left:

1946 HARLEY-DAVIDSON MODEL U 1200CC

UNITED STATES

Powered by the distinctive Harley "flathead" engine and produced in limited quantities until 1948, when the flathead gave way to the Harley "panhead" engine.

1946 IZH-350 350CC

USSR

Closely modeled after the German DKW cycle, whose facilities were seized by conquering Soviet troops at the end of World War II.

1947 INDIAN CHIEF 1200CC

UNITED STATES

Sporting the iconic skirted fenders that distinguish Indian from any other cycle ever produced.

Opposite: 1946 ***Harley-Davidson Model U***

HARLEY-DAVIDSON

MADE IN
Velocette
ENGLAND

Clockwise, from top left:

c.1947 ABERDALE AUTOCYCLE 98CC

UNITED KINGDOM

With its economical Villiers Junior de Luxe engine, this rare Aberdale had a top speed of only 34mph but reportedly achieved an astonishing fuel efficiency of almost 150 miles per gallon.

1948 VELOCETTE KSS MKII 349CC

UNITED KINGDOM

This magnificent specimen of the post-war Velocette was discovered in a shuttered bike shop in San Jose, California, after which it was deconstructed to the bare frame and restored component by component as if suspended in time.

1950 BSA COMPETITION ALLOY 350CC

UNITED KINGDOM

Defunct now for almost five decades, BSA once reigned supreme as the largest motorcycle company in the world, with stylish speed merchants like this.

Opposite: 1948 ***Velocette KSS MKII***

Clockwise, from top left:

1950 INDIAN SCOUT 250 440CC

UNITED STATES

This Indian Scout proudly displays its original factory paint and chrome, which reportedly have never been cosmetically restored; 1950 was the last model year for the Scout, and three years later, in 1953, Indian ceased all manufacturing activities.

1950 LA FRANÇAISE DIAMANT 125CC

FRANCE

Outfitted exactly as it was throughout its career as a fire-fighting vehicle at the small Aeroport de Montélimar in France, with side-mounted fire extinguisher and sand bucket, warning lights and horn, and a license plate that clearly identified it as a fire fighter with the single word "Incendie."

1952 MATCHLESS G3L 350CC

UNITED KINGDOM

Dating back to 1899, the Matchless marque had an enviable record of racing success, capturing the first-ever single-cylinder race at the prestigious Isle of Man Tourist Trophy in 1907 at an average speed of just over 38mph.

Opposite: 1950 ***La Française Diamant***

MINIMAX
EXTINCTEUR
PRENDRE
FRAPPER
ARROSER
AFHAKEN
INSLAAN

Clockwise, from top left:

1953 HOREX REGINA 350CC

GERMANY

Only a handful of years after the devastation of World War II, Horex brought the Regina 350cc to market, with telescopic forks, alloy drum brakes, and handsome lines featuring a fully enclosed drive train.

1954 GNOME ET RHÔNE LC531 175CC

FRANCE

Leg shield and running boards distinguish this fine example of the relatively rare Gnome et Rhône motorcycle, a marque more widely known in the aviation industry for its high-quality aircraft engines built during World War I and again during World War II.

1954 BSA B31 348CC

UNITED KINGDOM

Carrying the scars and stickers of a ride around the world in circa 1974 by the adventurer Stan Soames, this BSA registers 92,154 miles on its odometer and is basically untouched after its odyssey.

1955 **PUCH SGS** 250CC

AUSTRIA

Puch motorcycles are noted for their innovative "split single" two-stroke engine, featuring tandem cylinders with pistons driven by a single connecting rod.

1955 MOTOBI B200 SPRING LASTING 200CC

ITALY

Its egg-shaped engine with horizontal cylinders was an iconic feature for Motobi, a firm set up when Giovanni Bentley broke away from the family firm Benelli, only to rejoin it again in 1962.

Clockwise, from top left:

1956 JAMES COMET L1 98CC

UNITED KINGDOM

Only 3,000 miles from new on this unrestored James Comet L1 powered by a Villiers engine and sporting a vented leg shield.

c.1956 LAMBRETTA MODEL D 125CC

ITALY

A beautifully restored and stylish Lambretta scooter manufactured by Italian industrial conglomerate Innocenti; production of the Model D ceased in 1956, and few survivors exist today.

1957 NSU SUPERMAX S-20 SCRAMBLER 250CC

GERMANY

One of a limited edition of only about 230 Supermax Scramblers built by NSU for off-road racing in the United States at the urging of the importer and AMA Hall of Fame racer Earl Flanders.

Clockwise, from top left:

1960 MOTO GUZZI GALLETTO 192CC

ITALY

When it ventured into scooters in the 1950s, famed Italian motorcycle builder Moto Guzzi produced more of a hybrid than a classic scooter, incorporating large 17-inch wheels, a respectably sized and centrally mounted 192cc engine, double-seating, and classic design features.

1960 HARLEY-DAVIDSON TOPPER 165CC

UNITED STATES

The only scooter ever produced by Harley-Davidson, the Topper featured a rope-recoil starter similar to a lawnmower that would make chopper purists wince.

c.1961 BATAVUS G50 COMBISPORT 50CC

NETHERLANDS

Featuring its distinctive oval headlamp, bulge gas tank, and unusual front fork configuration, this handsome version of the Combisport shows barely any evidence of its age.

1963 NORTON ATLAS

750CC

UNITED KINGDOM

Bridging the period between the legendary Manx Norton racers and the Norton Commando, the Atlas 750 was designed to appeal to the U.S. market, with features such as high-rise handlebars.

Clockwise, from top left:

c.1963 TRIUMPH TIGER CUB 200CC

UNITED KINGDOM

An off-road version of the Tiger Cub, improvised with parts from multiple years, including a frame and other components reportedly produced by the famous racer Sammy Miller.

1965 BMW R50/2 500CC

GERMANY

The uncommon Dover white version of the BMW line of cycles, a radical departure from its signature high gloss black with white pinstripes.

1966 HARLEY-DAVIDSON AERMACCHI SPRINT H 246CC

UNITED STATES

A rare overseas venture for the Milwaukee-based American icon, importing this street scrambler by the Italian firm Aermacchi and then adding its own Harley badge.

Opposite: 1965 **BMW R50/2**

0811
BLITZ
Wartungsfrei
6 Volt

Clockwise, from top left:

1968 PANNÓNIA 246CC T5 246CC

HUNGARY

A "barn find" in 2016, this survivor of Hungary's last and most successful motorcycle firm was treated to a head-to-toe restoration in order to recapture its original glorious appearance.

1970 HONDA DAX ST70 72CC

JAPAN

This bantamweight packs a host of distinctive features, such as a pressed-steel "T-bone" frame, elevated handlebars, and a stylish perforated exhaust.

c.1970 EGLI-VINCENT 1150CC

SWITZERLAND AND ENGLAND

The renowned Swiss-designed Egli frame wrapped around a Vincent Series-C Black Shadow engine to create a near mythic blend of raw power and handling grace.

1971 BSA SCRAMBLER 250CC

UNITED KINGDOM

BSA traces its lineage to the Birmingham Small Arms Company BSA, founded in 1861 by fourteen gunsmiths to supply arms to the British government during the Crimean War.

Clockwise, from top left:

1973 ZÜNDAPP BERGSTEIGER C50 SCOOTER 50CC

GERMANY

The Bergsteiger, or mountain climber, offered simple and economical transportation but with surprisingly good hill-climbing ability for its 50cc engine.

1974 ZÜNDAPP BERGSTEIGER M50 SCOOTER 50CC

GERMANY

Unusual white sidewall tires accent this Bergsteiger, or mountain climber, noted for its surprisingly good hill-climbing ability.

1974 TRIUMPH BONNEVILLE T140 750CC

UNITED KINGDOM

Named after the Bonneville Salt Flats of Utah, a favorite venue for the hardy souls who chase elusive speed records.

Opposite: 1974 ***Triumph Bonneville T140***

TRIUMPH

MOTO GUZZI
FUEL
NGK

Clockwise, from top left:

1977 BENELLI 750CC SEI 750CC

ITALY

The Benelli Sei "Six" was the world's first production six-cylinder engine, developed under the stewardship of Argentinian industrialist Alejandro De Tomaso, who took control of Benelli in 1971.

1982 HONDA MOTRA CT50 49CC

JAPAN

Produced only from 1982 to1983, the Motra swore off flashy chrome in favor of a rugged military look with steel panels and wheels colored only in yellow or green.

1984 MOTO GUZZI LE MANS III CAFÉ RACER 850CC

ITALY

Italian flair accents a café racer with characteristic low handlebars, impressive seat cowling, and bulbous gas tank with knee-grip indentations.

Opposite: 1984 ***Moto Guzzi Le Mans III Café Racer***

Clockwise, from top left:

1998 MV AGUSTA F4 SERIE ORO 750CC

ITALY

Left untouched and unridden in its original crate, Number "8" in a series of only 300 is believed to be the first Serie Oro acquired by anyone other than a member of a royal family or an MV Agusta executive.

1999 HARLEY-DAVIDSON MT500 500CC

UNITED STATES

The military MT500, with its Rotax engine known for dirt bike racing, is one of the rarest Harleys ever produced; this one is #237 of only 355 produced in 1999.

1999 EXCELSIOR-HENDERSON SUPER X 1386CC

UNITED STATES

In the late 1990s, the Hanlon brothers of Belle Plaine, Minnesota, sought to resurrect the pre-Depression glory days of the Excelsior and Henderson marques once owned by Ignaz Schwinn of Schwinn bicycle fame. Despite an estimated $100 million of financing and nostalgic design features such as through-the-fender tubes, the modern-day Excelsior-Henderson met a similar fate to its defunct predecessors after fewer than 2,000 units were produced.

Opposite: 1999 ***Harley-Davidson MT500***

HARLEY—
DAVIDSON

2000 **"CAPTAIN AMERICA" BY PANZER** 1500CC

UNITED STATES

One of the reportedly 50 replicas produced by Panzer Motorcycle Company, the #27 is the most famous cycle of all time, named "Captain America" and ridden by Peter Fonda in the 1969 film classic *Easy Rider*.

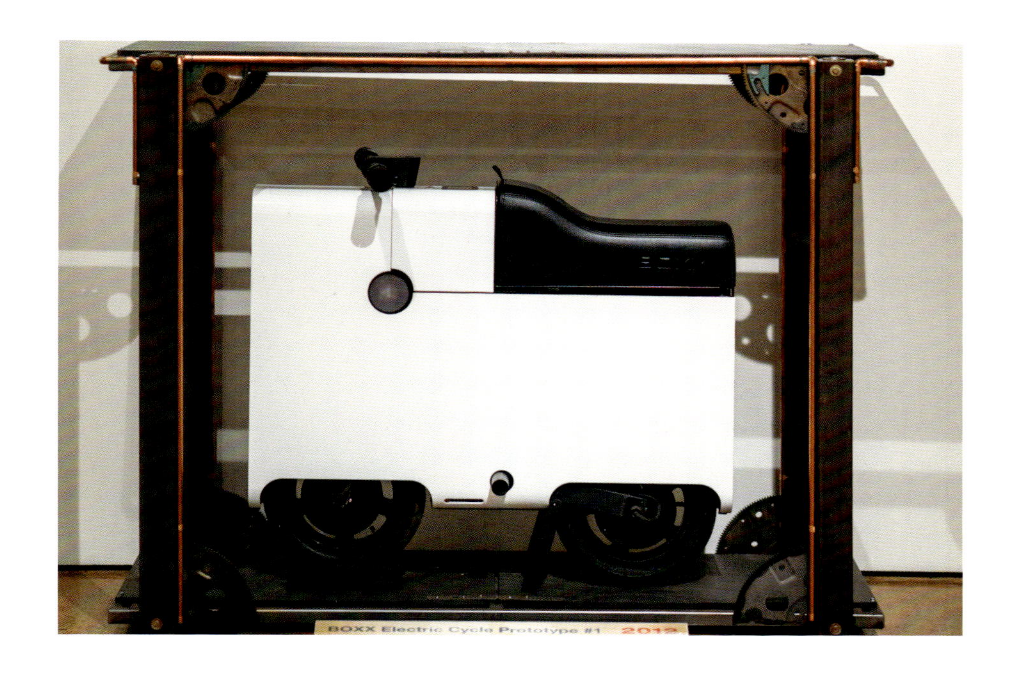

Clockwise, from top left:

2012 BOXX ELECTRIC CYCLE PROTOTYPE #1

UNITED STATES

The prototype of the lithium-powered, electric BOXX cycle, only 39.4 inches in length but long on innovative design and engineering excellence.

2015 INDIAN SCOUT 1133CC

UNITED STATES

A flawless example of Indian styling delivered straight from the dealer to the Haas Moto Collection, with no mileage ever recorded on the street.

2016 VICTORY EMPULSE TT

UNITED STATES

This limited-edition, electric-powered cycle instantly qualified as a rare collector's item after Victory abruptly announced cessation of all production activities in 2017.

2019 BROUGH SUPERIOR PENDINE SAND RACER 997CC

FRANCE

British in heritage but resurrected in a factory in the South of France, this timeless Brough Superior Pendine Sand Racer recaptures the century-old mystique of Brough Superiors that would race on a seven-mile stretch of beach on the south coast of Wales in the Welsh TT championships.

SIDECAR ALCOVE

Bridging the transition between the Race Track and the Custom Shop, the area known as Sidecar Alcove is a tribute to the segment of motorcycle culture where three-wheeled machines afforded a spacious compartment for companions. Whether intended for the rigors of wartime or the comforts of a country road, sidecars are a rich diversion from the mainstay of solo machines. Sidecar Alcove is populated with representatives from Russia, Germany, the Czech Republic, France, Denmark, Japan, and Italy—evidence that the unique desire and demands for sidecar transport represent a universal theme.

c.1917 NEW IMPERIAL RUSSIAN WAR PRODUCT 964CC

UNITED KINGDOM

Bearing the designation "War Product" on its engine, this New Imperial combination was produced by Great Britain during World War I for the Russian government, but was never delivered after the Bolshevik Revolution in Russia led the UK government to cancel the contract.

Clockwise, from top left:

1927 TERROT NS SPORT & BERNADET SIDECAR 500CC
FRANCE

The luxurious Bernadet sidecar found itself attached to stylish French and German motorcycles of the 1920s and 1930s, such as this 500cc four-stroke JAP-powered Terrot.

c.1928 PRAGA BD & FAVORIT SIDECAR 500CC
CZECH REPUBLIC

Defying its ninety-year-old provenance, this rare and handsomely restored combination has its sidecar mounted on the left, as was the Czech custom of this era.

c.1937 BROUGH SUPERIOR PETROL TUBE SIDECAR
UNITED KINGDOM

A heavily corroded "barn find" bears the scars of more than eight decades of neglect, but with its dignity and regal lineage intact; ingenuously designed with a pressurized petrol tube that looped around the sidecar and supplied fuel to a companion Brough Superior motorcycle.

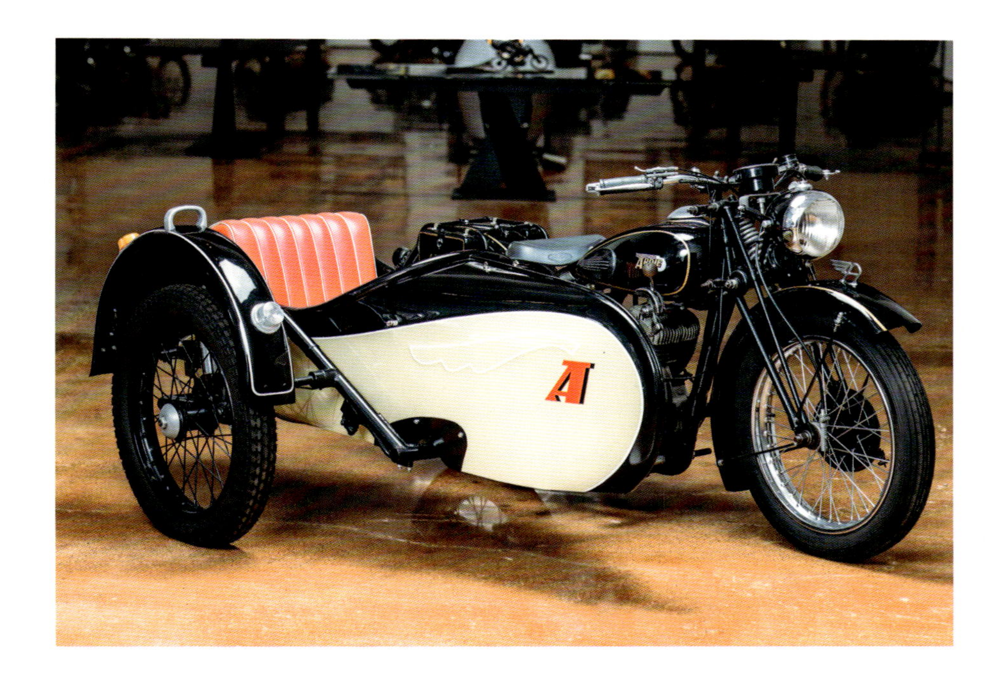

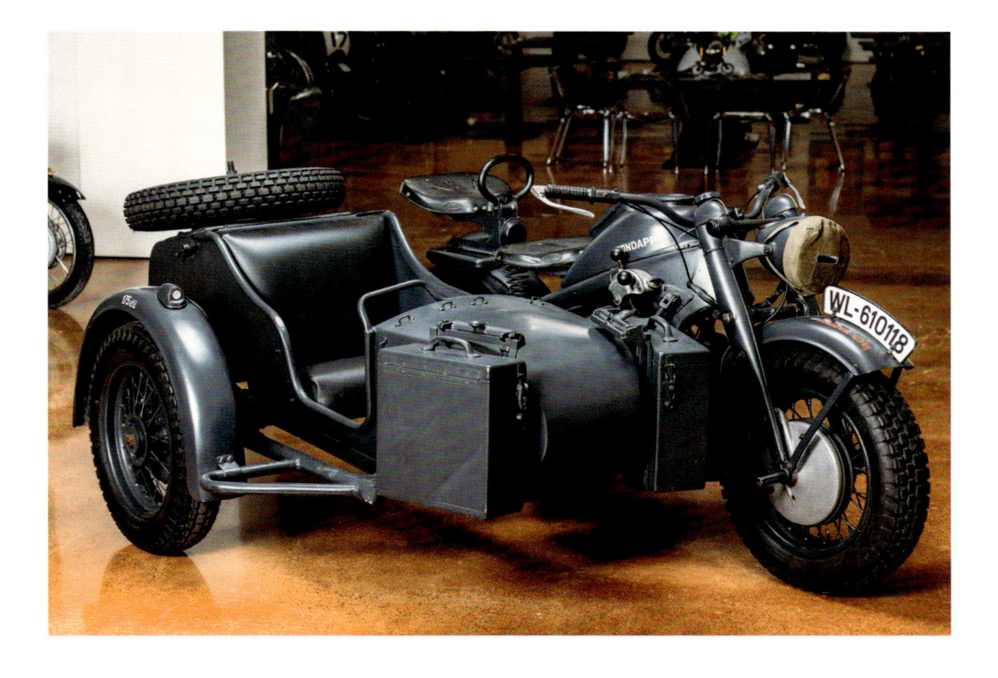

Clockwise, from top left:

1935/1958 **ARDIE RBU & STOYE SIDECAR** 500CC

GERMANY

The Stoye sidecar that complements this Ardie motorcycle was manufactured in 1958, more than two decades after the companion cycle; ironically, 1958 was also the last year of production for Ardie.

1943 **ZÜNDAPP KS750 & STEIB SIDECAR** 750CC

GERMANY

Developed during World War II, the rugged Zündapp KS750 with Steib sidecar saw action on every battlefront of the war, from the broiling sands of the Sahara to the frozen tundra of Russia.

1955 **ISO SCOOTER & SIDECAR** 125CC

ITALY

A rarity in almost every respect: a 125cc Italian scooter from the '50s tugging a factory-installed ISO sidecar; energized by a dual-piston, single-cylinder engine; and restored to recapture its impeccable period styling.

Clockwise, from top left:

1956 NIMBUS MODEL C & BENDER SIDECAR 746CC

DENMARK

Nicknamed the "Bumblebee" for its distinctive exhaust sound, the Model C featured an inline four-cylinder engine and shaft final drive; it often emerged from the factory with an attached sidecar such as this Bender.

1965 MARUSHO LILAC R92 & STEIB SIDECAR 500CC

JAPAN

True to its aesthetic name, this beautifully restored and ultra-rare Marusho Lilac R92 flat twin is joined with a German Steib sidecar in a highly unusual combination.

1969 HARLEY-DAVIDSON SERVI-CAR 750CC

UNITED STATES

Designed during the Great Depression, this rugged three-wheeler was produced from 1932 to 1975 as a service vehicle for auto dealers, small vendors, and even police departments.

2010 URAL PATROL 750CC

RUSSIA

A bevy of sparkling options and its eggshell patina camouflage this Ural's drab origins as a World War II army vehicle produced in the Ural Mountains of Russia, out of range of Germany's Luftwaffe bombers.

PART 2

THE

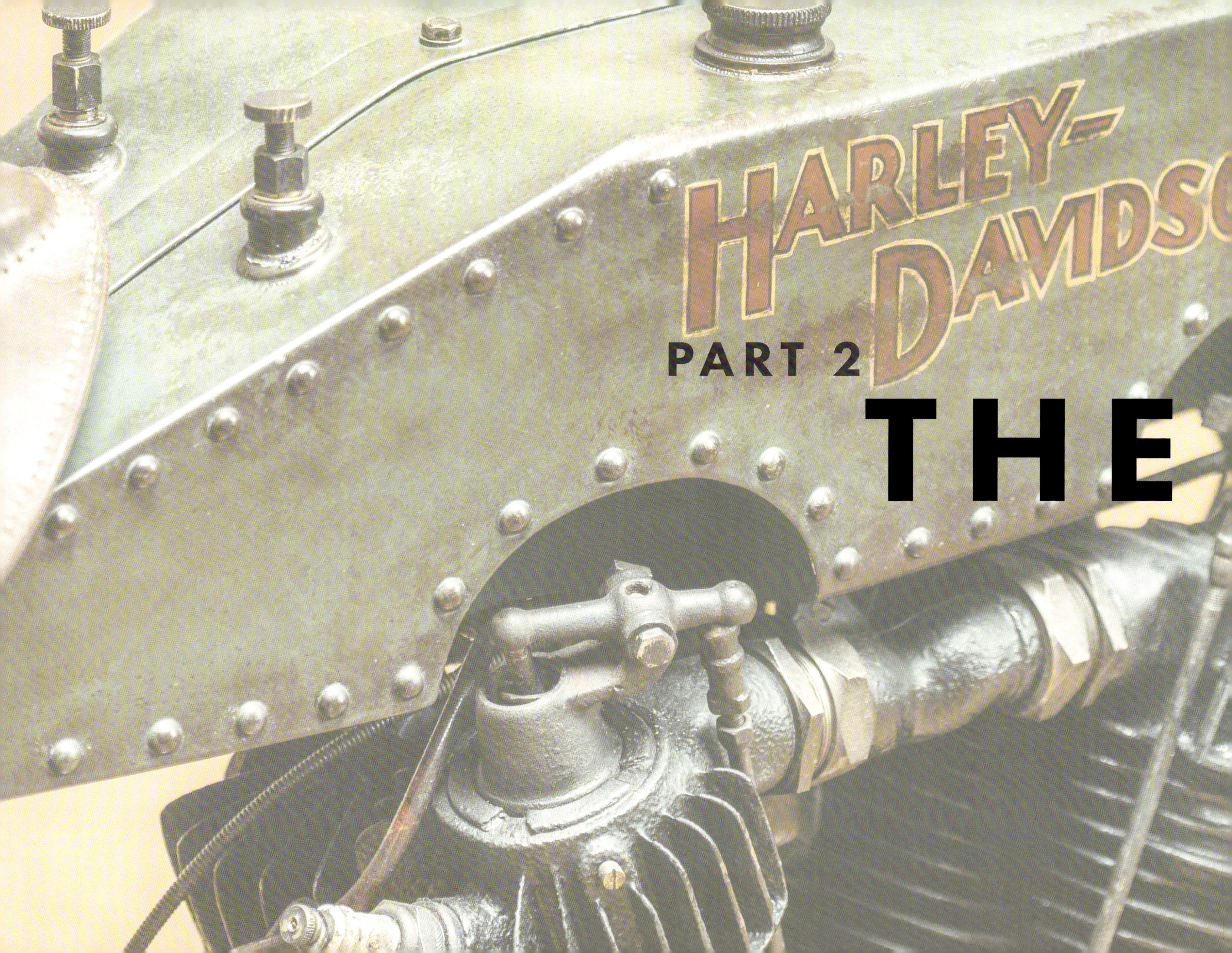

RACE TRACK

Yellow
47

THE COLD SCORCH OF FEAR

"Doubtless some readers will have been puzzled as to why a young man put his life at such unnecessary risk in the first place.... But young people need to test themselves. In domesticated societies so bereft of wildness, they need to register the cold scorch of fear now and then in order to feel truly alive."

–Tim Winton

A few years ago, when I was attempting to pilot a Harley for the first time, learning that the "friction zone" refers to the process of engaging the clutch and has absolutely nothing to do with marital discord, I came across an article in the *New York Times* that would forever shed light on why I seem to be drawn to one risky enterprise after another. It was an op-ed piece by the acclaimed Australian author Tim Winton, sharing his thoughts on the recent death of Nicholas Mevoli, a world-renowned free-diver who perished at the age of thirty-two during a diving competition at Dean's Blue Hole in the Bahamas. Winton's comments are to the left.

"The cold scorch of fear ... in order to feel truly alive." Winton sure got it right, all except for his limiting this to "young" people. I'm no longer young and haven't been for several decades. Yet the older I become, the more I crave *"the cold scorch of fear now and then in order to feel truly alive."*

I tossed my acrophobia overboard at the age of fifty-five to partake of the wind-whipped thrill of leaning out of an open helicopter with camera in hand in order to capture aerial images for *National Geographic.* My fear of heights was nowhere to be found when I was tethered to a different type of "chopper" by nothing more substantial than a leather harness. And I shuttered my fear of extreme cold inside a closet when, as part of a ten-year stint with *Nat Geo,* I flew over the Arctic for four years to capture aerial images of a part of the world that has come under assault from climate change.

And then I fired up my first Harley a decade later at the age of sixty-four. The cold scorch of fear was back again—just as tart, just as addictive the second time around. Anyone who tosses a leg over this ground-based type of chopper and does not confess that the danger is part of the thrill is not telling you the truth, the whole truth, and nothing but the truth. There are plenty of incidents that you witness yourself and plenty of anecdotes that you hear from others that raise a whiff of fear in even the most battle-hardened biker.

But the fear need not be a physical fear in order to raise the stakes of the endeavor ... the cold scorch of *failure* is close enough to physical danger to be a worthy twin brother. When I left a comfy law partnership at the age of thirty-one and launched myself into the bracing waters of private equity, I had no business (and no credentials) for achieving success, but the outsized rewards for "first movers" in the field simply proved irresistible. And the likelihood of failure was clearly part of the allure.

3216
BONNEVILLE NATIONAL
2016
INSPECTED
SPEED WEEK
INSPECTED
BONNEVILLE
SPEED WEEK
69 ANNUAL
EAGLE

So too when I started collecting vintage and custom cycles barely three years ago. An audacious endeavor for a newbie who could not even explain at the time how a carburetor works (I might fumble a bit even today if you ask me exactly how it works, 140 cycles later).

It was the likelihood of failure, the improbability of success, that drew me into the tornado of private equity, and then the hallowed halls of *National Geographic*, and now the world of exhibiting motorcycles inside an edifice that I dare to call a *museum*. For when you cobble the word *museum* onto the outside of your building, you have taken on an awesome responsibility to educate, to inspire, and to distinguish yourself from others ... and you can easily detect the cold scorch of fear in the back of your throat.

—Bobby Haas, 2017

Arranged in an oval representative of the early board tracks, the Race Track exudes raw speed, with a palpable sense of the few souls hardy enough to take flight at great risk to life and limb. Spanning more than a hundred years of racing prowess, this exposure to the culture of motorcycle racing offers a distinct glimpse into a world where the ghosts of the past, such as those who straddled a 1914 Indian Board Tracker, and the speed merchants of modern times, such as the ones who challenge the Bonneville Salt Flats, mingle together as members of one continuous breed of man-and-machine. Every cycle is a magnificent and often completely unique specimen of moto racing, as the designers tinkered with frames and engines to coax the maximum speed and endurance out of their steeds.

1904 PEUGEOT MODEL D RACER 330CC

FRANCE

One of very few survivors of this early 20th-century model, this bike is presented in virtually untouched condition, with saddle and handlebars in the lowered racing position and its chassis stripped down to the bare essentials.

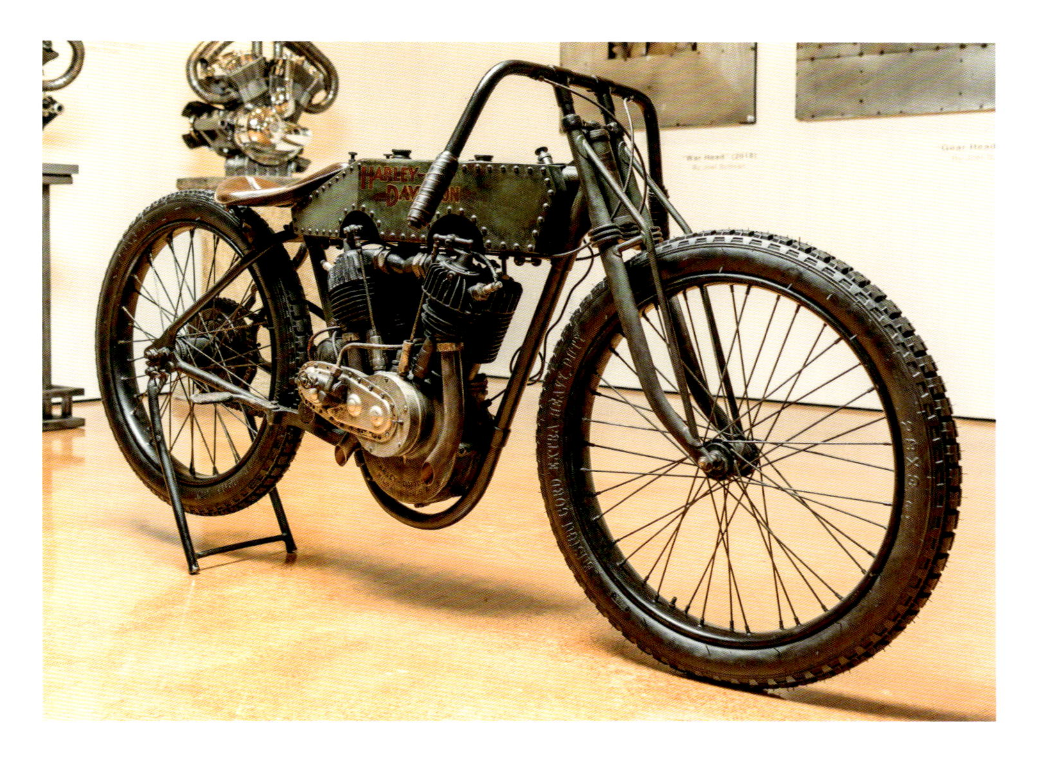

Clockwise, from top left:

1914 INDIAN BOARD TRACK RACER 1000CC

UNITED STATES

Restored to its century-old luster with a 1000cc V-twin that bears an uncanny resemblance to its more modern descendants.

1915 HARLEY-DAVIDSON BOARD TRACK RACER MODEL 11-K

1000CC UNITED STATES

This rare and exquisitely restored example of the 11-K board track racer that marked Harley's successful entrance into the world of board track racing more than a hundred years ago.

c.1926 TERROT HT SPORT 350CC

FRANCE

With a powered motorbike tradition dating back to 1902, Terrot's Dijon factory first built motorcycles with engines supplied by proprietary engine manufacturers such as the English firm JA Prestwich (JAP). In the mid-1920s, however, it finally produced its own engines for the HT line of Terrot cycles. This HT Sport, with flat handlebars and number plaque, was noticeably outfitted for racing.

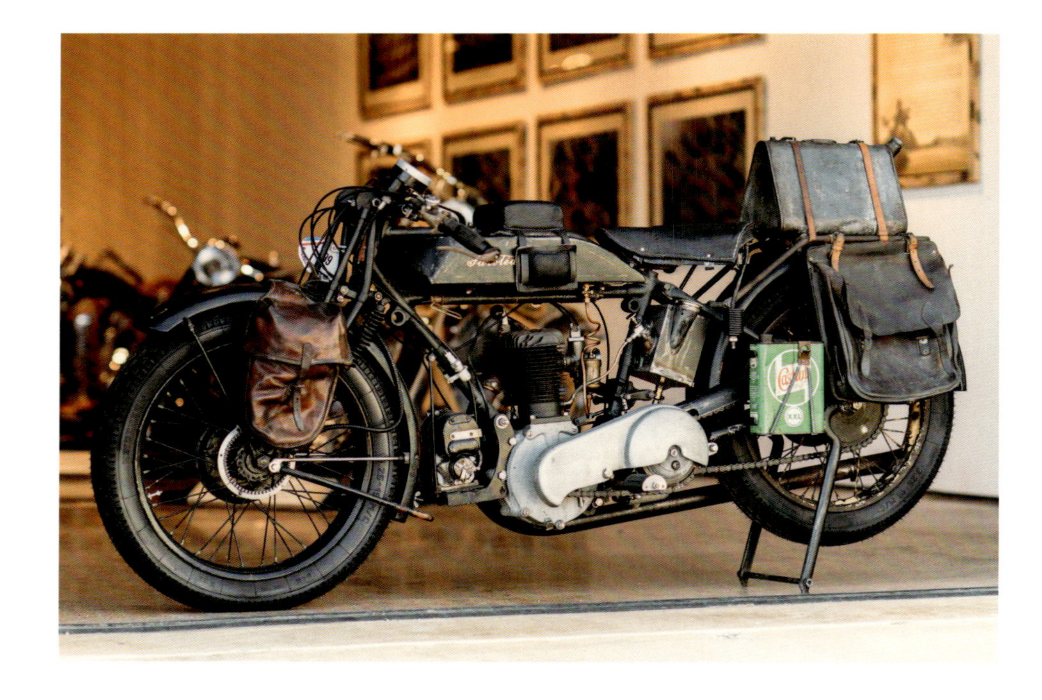

Clockwise, from top left

c.1928 **FN M67A FACTORY RACER** 500CC

BELGIUM

With its bulbous, large-capacity gas tank, this model by Fabrique Nationale was ideal for long distance races and is believed to have competed as recently as the 2016 Austrian "hand shifter run."

1928 **SAROLÉA MODEL 23T SUPERSPORT** 500CC

BELGIUM

Equipped for the rigors of a long distance ride, complete with an extra tank of gas.

c.1929 **GILLET HERSTAL SUPERSPORT COMPETITION** 500CC

BELGIUM

Proudly sporting the bumps and bruises of a distinguished racing career on the European circuit.

Opposite: 1929 ***Gillet Herstal Supersport Competition***

Gillet
HERSTAL

1930 MONET-GOYON SUPERSPORT MODEL H RACER 500CC

FRANCE

Bearing a striking resemblance to the Belgian Gillet Herstal, the two might actually have met on the track in the 1930s.

Clockwise, from top left:

1951 TRIUMPH BONNEVILLE SALT FLATS RACER 649CC

UNITED KINGDOM

The historic Triumph racer clocked at 122mph in 1953 at the Bonneville Salt Flats with thriteen-year-old Bobby Sirkegian in the saddle; restored by Sirkegian himself after over fifty years in his family.

1951 HARLEY-DAVIDSON WR RACER 750CC

UNITED STATES

Produced specifically as a factory racer to challenge the Scout V-twin of archrival Indian in competitions on oval track speedways.

c.1954 ADLER RS250 250CC

GERMANY

Capable of coaxing an astonishing speed of almost 120mph out its small 250cc engine with radiator cooling.

Clockwise, from top left:

1955 CECCATO CORSA 75CC

ITALY

Barely known outside its native Italy, the Ceccato line of small displacement racers featured this single-overhead-camshaft Corsa, whose lighter weight was particularly well suited for long distance road races.

1956 NSU SPORTMAX 250 250CC

GERMANY

One of only thirty-six produced by NSU from 1954 to 1956, the Sportmax was capable of achieving 130mph; expertly restored by George Beale of Great Britain.

c.1956 DUCATI GRAN SPORT 98CC

ITALY

The Gran Sport became a mainstay of Ducati's much heralded racing prowess, with an engine designed by the legendary Fabio Taglioni; note also the horn and lights required by Italian regulations for public road, long distance events.

1958 DUCATI TRIALBERO DESMO 125CC

ITALY

The design of this highly innovative triple-overhead-camshaft racer by Fabio Taglioni marked a defining moment in Ducati history and motorcycle engineering.

c.1959 MOTO MORINI 250CC

ITALY

This restored 250cc cycle is representative of a distinguished line of small displacement Italian racers.

Clockwise, from top left:

1960 THE PERIL SPEED EQUIPE "YELLOW PERIL" 650CC

UNITED KINGDOM

The first of the Peril Speed Equipe sprinters designed, built, and raced by British carpenter/cabinetmaker Bill Bragg, who bent the forward-facing exhaust pipes inthe grill of the gutter outside his home. Attached to a sidecar, the Yellow Peril set a world speed record of 147mph.

1961 THE PERIL SPEED EQUIPE "SCARLET PERIL" 650CC

UNITED KINGDOM

One of the famous trio of sprinters by Bill Bragg, this "kneeler" added swing arm rear suspension to cope with the bumpy tracks of this era.

1962 THE PERIL SPEED EQUIPE "BLUE PERIL" 650CC

UNITED KINGDOM

The Peril Speed Equipe racers disappeared for over thirty years following designer/builder Bragg's immigration to Australia in 1966, only to be discovered in 1999 in a collapsed shed outside the home of Bragg's racing rival Ron May. Afterwards, the three Perils were restored, occasionally raced, and eventually reunited at the London Motorcycle Museum to celebrate their unique place in racing history.

Clockwise, from top left:

1962 PETTY-MOLNAR NORTON MANX ROAD RACER 519CC
UNITED KINGDOM

Drenched in British racing tradition and a veteran of the North American circuit, this Manx features a distinctive frame developed by racing icon Ray Petty, as well as a Molnar short-stroke engine expertly crafted by the famed Bob Barker.

1966 RICKMAN TRIUMPH STREET MÉTISSE 649CC
UNITED KINGDOM

A futuristic front fairing highlights the first production road bike ever to feature a disc brake, from legendary British racers and designers Derek and Don Rickman.

1966 TRIUMPH STREET TRACKER SPECIAL 750CC
UNITED KINGDOM

An exquisite makeover of the T120 Bonneville by American flat-track racer Ron Peck, complete with Morgo cylinders, Ceriani forks, and Mikuni carbs.

1967/2007 SEELEY-TAIT CUSTOM RACER 492CC

UNITED KINGDOM

A testament to perseverance and ingenuity, this cycle's home-built three-cylinder engine was attached by designer Bob Tait to a Seeley Yamsel frame four decades later in 2007.

1968 **LYNTON 500** 500CC

UNITED KINGDOM

The only Lynton racer ever designed and built by the legendary Colin Lyster, this uniquely configured racer, fitted with a custom cylinder head atop a Hillman Imp engine, achieved widespread notoriety despite never having raced.

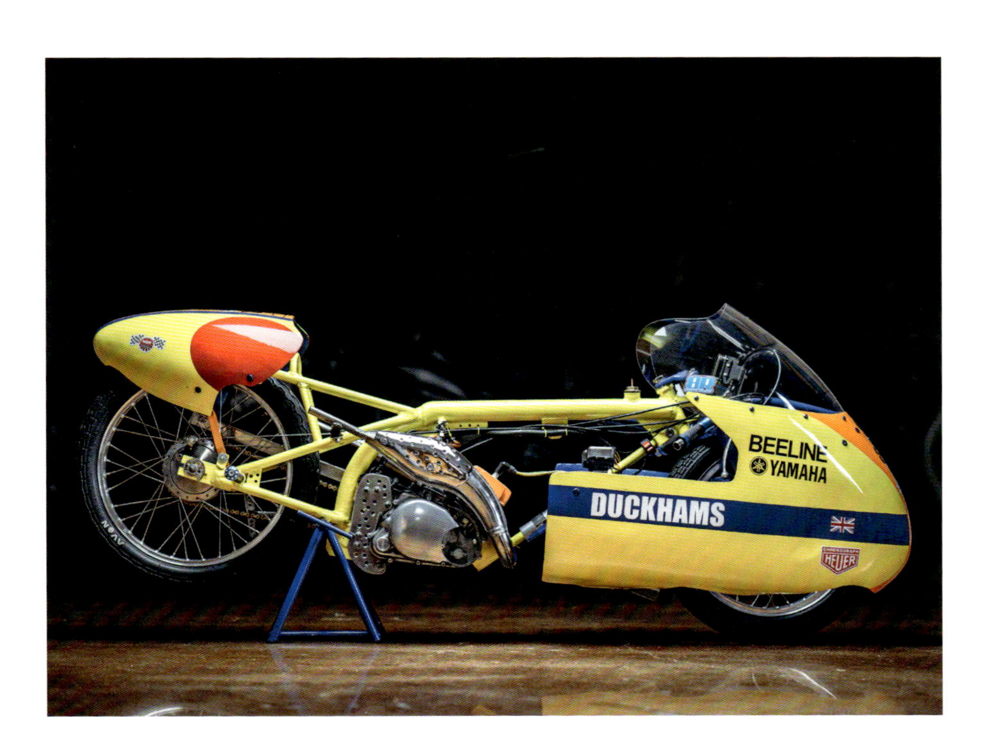

Clockwise, from top left:

c.1970 JIM GEE TRIUMPH TWIN-ENGINE RACING COMBINATION 1246CC

UNITED KINGDOM

Designed, built, and raced by Jim Gee, who developed this racer over a period of ten years, fashioning its unique power station in which the two engines rotated in opposite directions.

1970 KAWASAKI H1R 500CC

JAPAN

One of only about sixty factory racers produced by Kawasaki in 1970, this H1R was restored by Dave Crussell, one of the most respected Kawasaki restoration specialists ever.

c.1970 BEELINE YAMAHA SPRINTER 97CC

UNITED KINGDOM

Weighing in at a mere 110 pounds, this diminutive sprinter was a heavyweight in British racing history, garnering three world records over the period 1971 to1980. Fitted with a race kit, its circa 1970 Yamaha twin engine was capable of revving to an incredible 14,500 rpm.

1974

HUSQVARNA CR400 MOTO-CROSS 400CC
SWEDEN

Husqvarna's heritage included chainsaws and sewing machines, but it was the "Husky" moto-cross that gained dirt racer fame, capturing nine World Moto-Cross Championships, a like number of Baja 1000 enduros, and the allegiance of the "King of Cool" Steve McQueen, who rode a Husky 400 in the movie *On Any Sunday*.

Clockwise, from top left:

c.1975 **BULTACO PURSANG MK 8 MOTO-CROSS** 250CC

SPAIN

Restored to virtually mint condition, this Pursang MK 8 recalls the glory days of famed moto-cross marque Bultaco, a company formed when Francisco Xavier Bultó bolted from Montesa with virtually the entire racing department when Montesa opted to withdraw from the sport in the late 1950s.

1979 **MBA** 125CC

ITALY (ENGINE) AND FRANCE (FRAME)

French Grand Prix riders Jean Louis Guignabodet and Patrick Fernandez commissioned fellow countryman Eduard Morena to build the custom frame for this Morbidelli-Benelli-Armi.

1988 **DUCATI 851 SUPERBIKE TRICOLORE** 851CC

ITALY

Emblazoned with traditional Italian colors, this cycle is part of a limited edition of only 311 Ducati Strada Superbikes that helped trigger a resurgence of the Ducati brand.

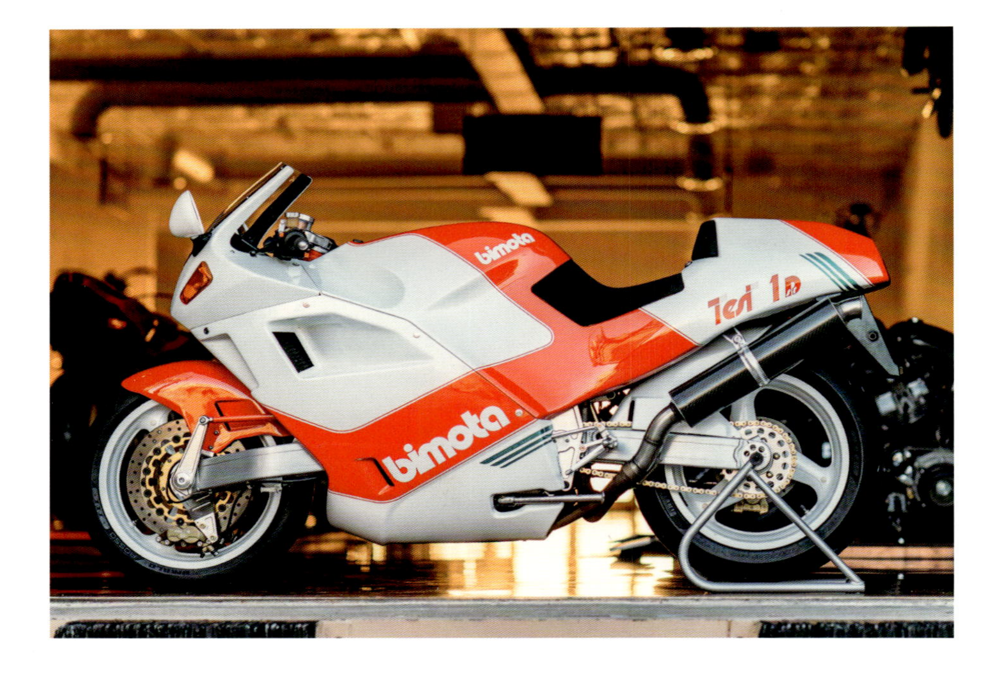

Clockwise, from top left:

1988 DUCATI PASO 748CC

ITALY

Produced in the same year as its Tricolore cousin, this albino version of the Ducati Paso trumpeted the conspicuous consumption of the *Bonfire of the Vanities* era.

1992 BIMOTA TESI 1DSR 904CC

ITALY

Founded in 1966 as a manufacturer of HVAC ductwork, Bimota gained fame in the 1970s as a designer of highly innovative, limited production motorcycles. This Bimota Tesi 1Dsr is one of only 144 ever built, with its signature hub-centric steering that separates the steering, braking, and suspension systems.

2017 SUTER MMX2 GRAND PRIX RACER 600CC

SWITZERLAND

After surviving the bone-jarring rigors of the 2017 Grand Prix circuit, this 600cc Moto2 racer was acquired by tThe Haas Moto Museum immediately before its final race in Valencia, Spain.

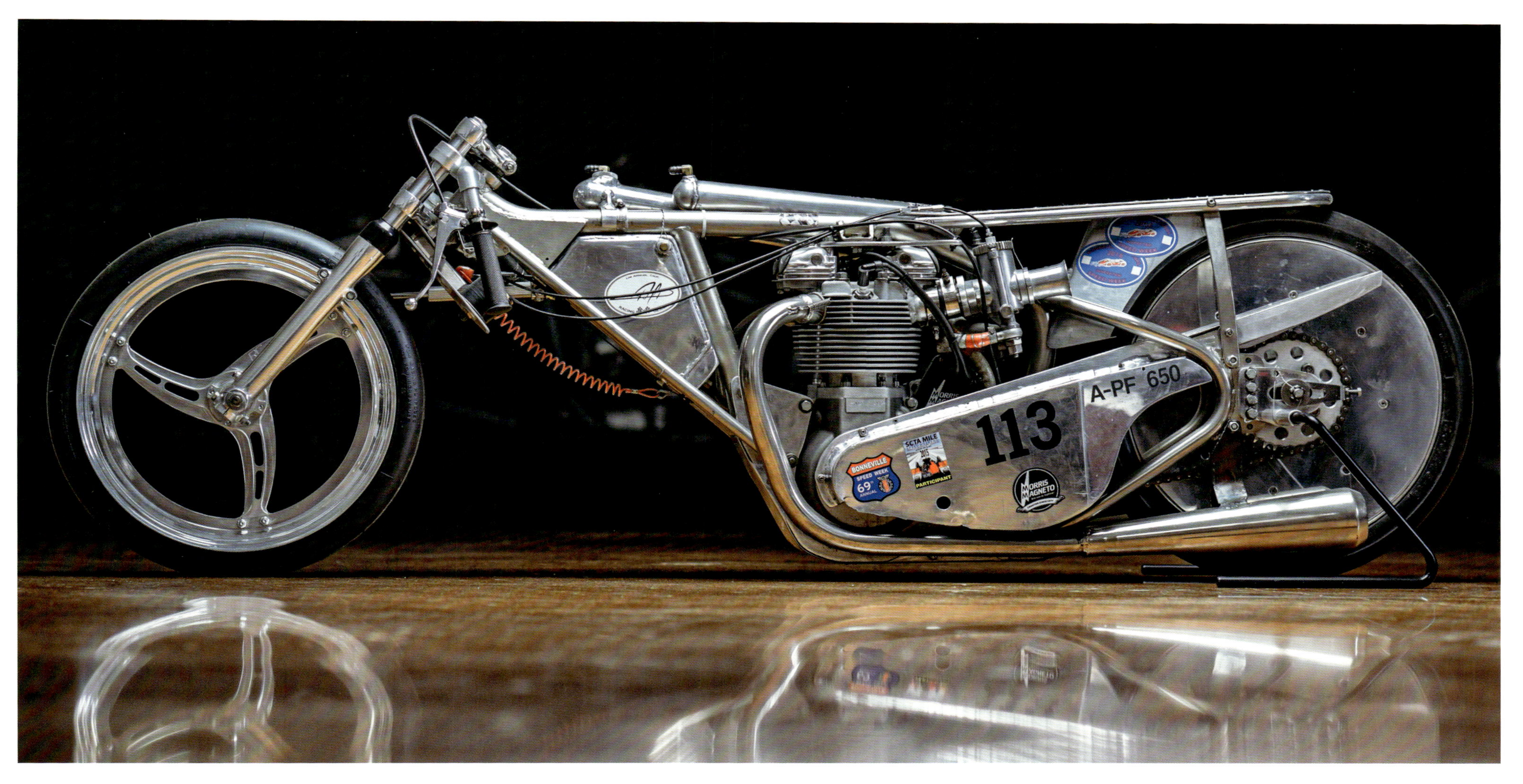

2014 ASYMMETRIC AERO 650CC

UNITED STATES

ALP RACING & DESIGN

With none of the wind-deflecting bodywork characteristic of land speed racers, the Asymmetric Aero by Alp Sungurtekin set eight world records, including 175.625mph at El Mirage Dry Lake, California, qualifying as the fastest unstreamlined, pushrod racer ever with an engine capacity under 1000cc. Its asymmetric handlebars allowed Alp to twist his shoulders perpendicular, not parallel, to the ground, reducing wind drag in this contorted position.

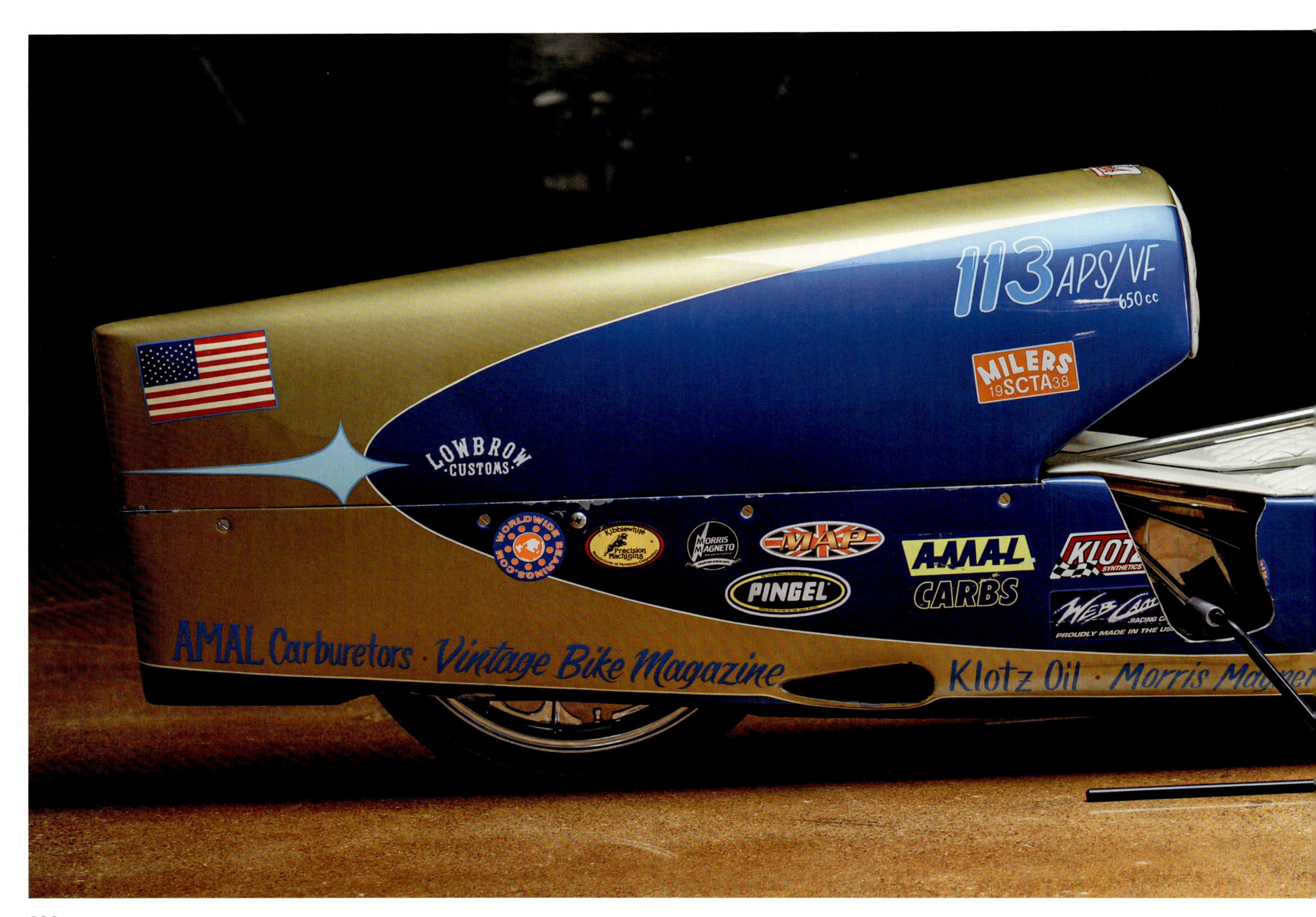
113 APS/VF
650 cc
MILERS
19 SCTA 38
LOWBROW
CUSTOMS
WORLDWIDE BEARINGS.COM
Kibblewhite
Precision Machining
MORRIS MAGNETO
MAP
PINGEL
CARBS
AMAL Carburetors · Vintage Bike Magazine
Klotz Oil

2017 **T200** 650CC

UNITED STATES

ALP RACING & DESIGN

Designed and built by Alp Sungurtekin around a 1950 Triumph 650cc twin-cylinder pushrod engine, the T200 set an SCTA two-way average land speed world record of 170.883mph at the Bonneville Salt Flats. Its partially streamlined, aluminum alloy bodywork transforms itself into a complete wind-deflecting capsule once its pilot Alp inserts himself into the body-hugging cavity between the fore and aft sections.

PART 3

THE CU

STOM SHOP

GRAND PRIX de SUISSE
2021

CHASING PERFECTION

Our documentary—named, aptly enough, *Leaving Tracks*—used to have the subtitle *Chasing Perfection*, but when we were designing the artwork for the streaming platforms, we convinced ourselves that the phrase *Chasing Perfection* was so powerful that it would fight for top billing with *Leaving Tracks*, so we dropped *Chasing Perfection* and left the title all by itself on the poster art. Besides, we figured the audience would realize soon enough that *Leaving Tracks* was itself a *double entendre* that refers not only to motorcycles "leaving tracks" in their wake but also how life challenges each of us to "leave tracks" when it is time for our wake.

Actually, the phrase *chasing perfection* is the title of an essay I wrote sixteen years ago for the first of five books that *National Geographic* published of my photographic work. There too, the phrase had two meanings—chasing the perfect photographic image and chasing perfection in our lives—as is apparent from the italicized excerpts below from that sixteen-year-old essay.

The synopsis of our film emphasizes how as founder of the Haas Moto Museum, I was bonded to our premier custom builders, in fact to all custom builders worldwide, by the frustrations of chasing perfection. (*The pursuit of perfection is a dangerous game, played on a field with endless distractions, hemmed in by blind alleys and cul-de-sacs.*)

Building a custom motorcycle is a chase for perfection, a journey littered with potholes even for the greatest of custom builders—the Hazans and the Rodsmiths and the Siegls and the Shinyas of this oil-stained world. But it is a chase that artists and athletes and parents pursue anyway. (*Relentlessly seeking perfection—in ourselves or others—sets a standard that resembles a much too narrow needle through which the course thread of human nature chafes against the eye and will not fit.*)

It is a chase that custom builders are drawn to, time and again, every time the first part for a new project is placed on a lift or in a vise. (*We all learn soon enough that the playing field of our lives is not a well-ordered rectangle; the rules of the game are seldom clear, or even fair. Perhaps that's one reason why the artist turns to art, in search of a blank canvas that offers up a brand new start, a four-cornered world where perfection is still a worthy goal.*)

Whether as custom builders or artists or parents or simply those who live so-called "regular" lives, we often return to the playing field and pursue the chase again. (*In the end, the chase itself might simply be a harmless way of raising the bar, of aspiring to levels we might not otherwise achieve. Those who chase this elusive goal do so with full knowledge that the quest will never be over ... but the journey never be pointless.*)

For everyone who is reading this, you have probably sought perfection in some aspect of your life and rarely, if ever, attained what you were seeking. (*But the joy is still yours. As I wrote back in 2004, "The search has no logical outcome, just the thrill of having played the game."*)

–***Bobby Haas***, 2021

KRGT-1S PROTOTYPE

2032CC (2016)

ARCH MOTORCYCLE

UNITED STATES

Affectionately known as "the Mule" by the artisan-engineers at Arch Motorcycle co-founded by Keanu Reeves and Gard Hollinger, the Mule served as the prototype for the 2018 KRGT-1s "Sport Cruiser" introduced at the EICMA motorcycle show in Milan in November 2017; the KRGT-1s sports a 124-cubic-inch, 2032cc V-twin engine, 17-inch carbon fiber wheels, and a host of other bespoke features.

ARCH
MOTORCYCLE
YOSHIMURA
ALPHA

BMW ALPHA

740CC (2017)

MARK ATKINSON AND MEHMET DORUK ERDEM

UNITED STATES

An alliance between strangers—Utah builder Mark Atkinson and Istanbul designer Mehmet Doruk Erdem—produced one of the most iconic cycles ever. With a streamlined body pressed "from whole cloth" here, layers of basalt lava and carbon fiber, Alpha is powered by a BMW K75 triple controlled by one-off hub-centric steering. Shifting is actuated by a stirrup around the left knee with a rear brake by the right. The dual kidney grills are a nod to the classic BMW 328 Mille Miglia racer. Great white shark elegance blends with turbocharged power to create literally one for the ages.

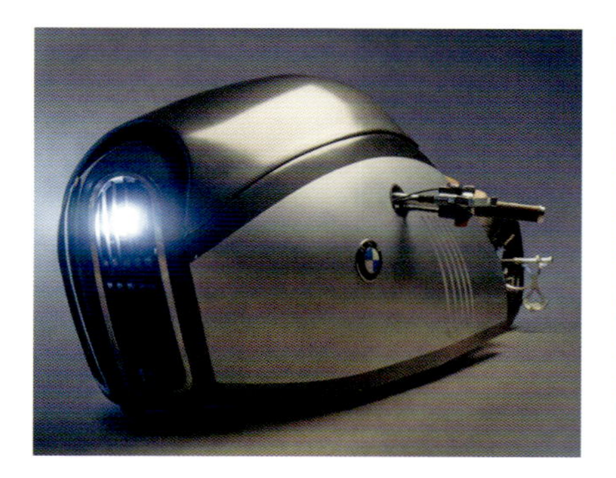

RACER-X

(2019)

MARK ATKINSON

UNITED STATES

Determined to design and build an e-bike from scratch that would radically depart from IC tradition, Mark Atkinson enclosed the internals with two sets of carbon fiber panels in a crossing X pattern. Adding hubless wheels and duo-pivot steering enabled the Racer-X to emerge with an arresting chassis that enclosed engineering excellence.

EVE MARK II

125CC (2016)

BANDIT9

VIETNAM

Its champagne patina celebrates the radical yet minimalist design of this customized creation from Vietnam, the first in a strictly limited series of only nine Eve Mark II motorcycles.

L•CONCEPT

125CC (2019)

BANDIT9 VIETNAM

Fashioned exclusively for the Haas Moto Museum, this ebony version of the limited edition "L • Concept" was designed and built by the avant-garde firm Bandit9 of Vietnam. Its suspended 125cc air-cooled, four-stroke engine is visually the center of gravity of a vehicle with such futuristic features as a teardrop saddle, uniquely shaped handlebars, sculpted grips and levers, and flat dual shocks.

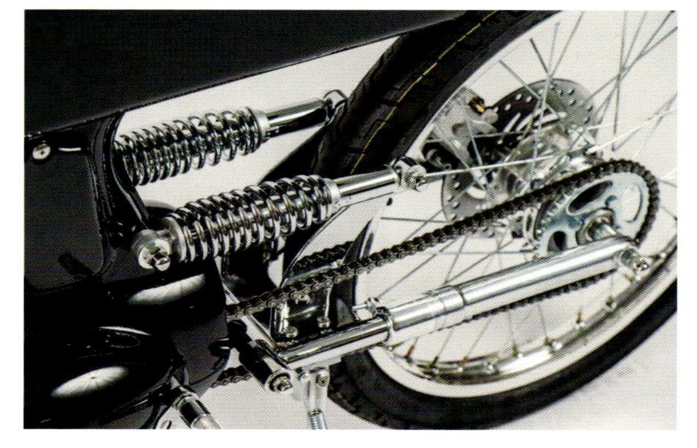

BUENO/HALLYDAY CUSTOM CYCLE

865CC (2018)

SERGE BUENO UNITED STATES

The only custom bike ever produced by moto aficionado Serge Bueno was designed and built for fellow countryman and rock legend Johnny Hallyday, known as the "Elvis Presley of France." Bueno added a bevy of custom features to the Triumph 865cc power station to create a muscular, racy, and vintage look that reflected the personality of the rock legend, who tragically died before ever taking possession of this stunning creation.

MOTO CENTURY

(2019)

CHABOTT ENGINEERING

UNITED STATES

Commissioned by the Haas Moto Museum to bring to life his dream project—a reimagined circa 1915 Indian cycle adapted to incorporate his signature style—Shinya Kimura labored for more than one year before the dream became reality. His almost archeological search for a period-correct Indian engine finally yielded results, enabling Kimura to create a seemingly "century-old" customized Indian with his fingerprints visible throughout.

Neko

NEKO

750CC (2019)

CHERRY'S COMPANY

JAPAN AND UNITED STATES

Master craftsman Kaichiroh Kurosu of Cherry's Company is noted for his alchemy in converting Harley-Davidsons into customized works of art that visualize the future through the lens of the past. The tight quarters of his shop on the outskirts of Tokyo is in the venue in which expansive ideas take shape, as with Neko, the Japanese word for "cat," which gives every indication that it is about to pounce onto the streets. Kurosu worked his magic on Neko by first slicing a 1972 Harley Shovelhead from two cylinders down to one and then customizing the frame and other elements, before wrapping this motorized predator in carbon fiber with gold accents.

AILERON

1375CC (2009)

DOTSON DESIGN CUSTOM CYCLE

UNITED STATES

Powered by a Kiwi Indian 84ci engine, this build by Christian Dotson continued his innovative styling and a host of custom features, such as front fork legs and rear swing arm fashioned from tapered tubing from early 1930s Ford automotive suspension; and a single downtube frame containing the engine oil. Like its older brother Swingshot, also designed and built by Dotson, Aileron was named "America's Most Beautiful Motorcycle" in 2009 by the Grand National Roadster Show.

SWINGSHOT

1525CC (2007)

DOTSON DESIGN CUSTOM CYCLE

UNITED STATES

Conceived and designed by Christian Dotson, Swingshot boasts a host of dynamic and innovative features, including radical forkless front design; hub-centric steering; twin hand-shaped fuel tanks and fender halves; adjustable automotive leaf springs; and a frame that serves as the receptacle for the engine oil and electrical wiring. Swingshot has earned a bevy of honors, including being named "America's Most Beautiful Motorcycle" in 2007 by the Grand National Roadster Show.

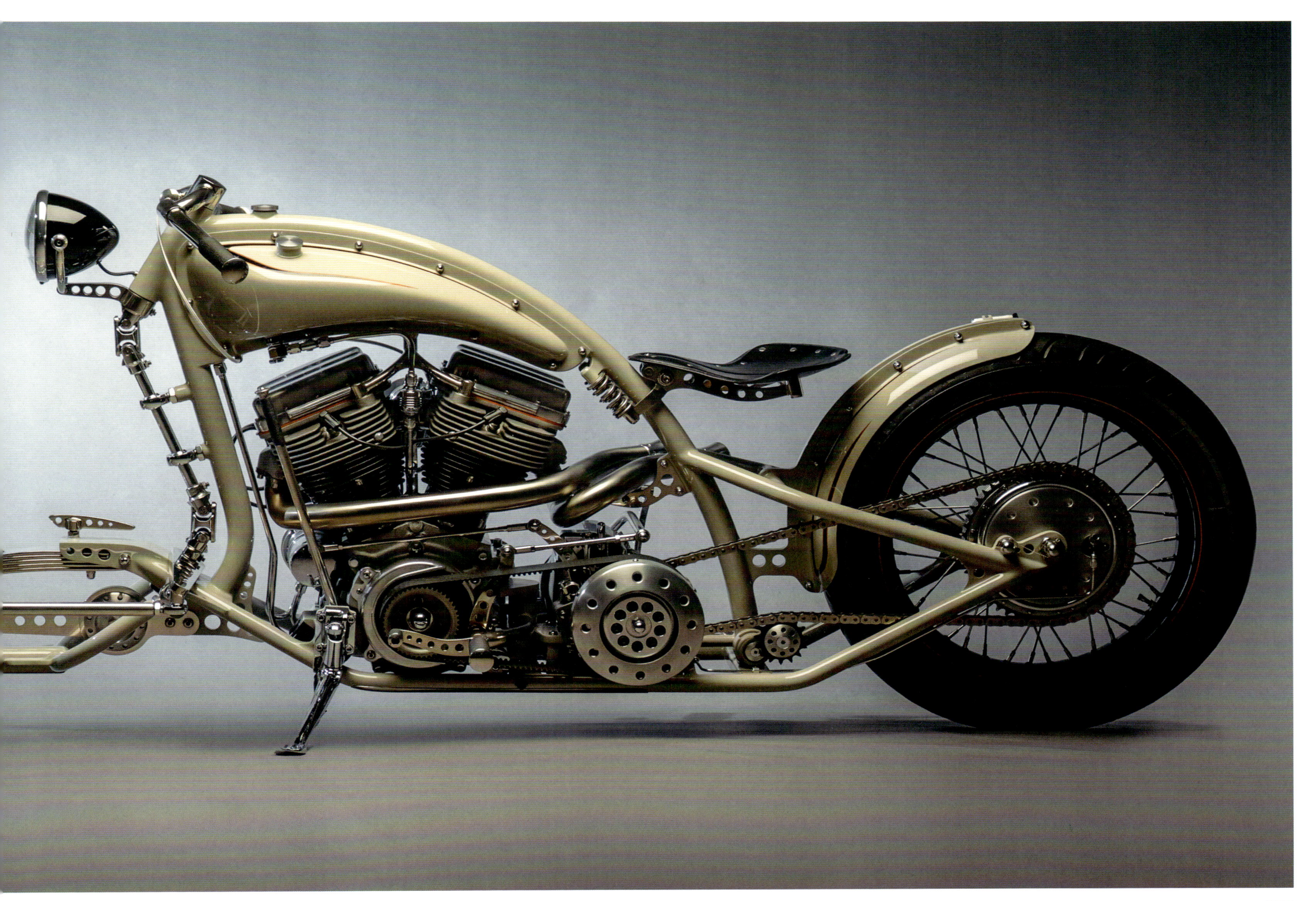

F131 HELLCAT

2147CC (2007)

CONFEDERATE

UNITED STATES

Poised for action but rarely engaged, this Hellcat has barely registered any mileage on its monstrous 2147cc engine.

R131 FIGHTER

2146CC (2012)

CONFEDERATE

UNITED STATES

Poised for action but rarely engaged, this limited-edition Confederate, one of only ten ever produced, registered just 270 miles on its monstrous 2146cc engine.

SEVEN

(2015)

JEREMY CUPP

UNITED STATES

Inspired by the famous 1934 CAC Harley-Davidson flat track racer, the seventh custom build by Jeremy Cupp fitted a 500cc Buell Blast engine with reverse-position Ducati cylinder heads coupled with a pre-1959 Triumph transmission; the thoroughly unique combination captured numerous custom titles, including the 2016 J&P Cycles Ultimate Builders Championship.

MADE IN USA

SX1250 DIRTSTER

(2018) JEREMY CUPP

UNITED STATES

With S&S 1250cc V-twin cylinders and Buell flywheels and heads, this dirt-ready Sportster is enlivened with liberal amounts of carbon fiber wrapping. Its knobby tires, naked rear wheel, and high chopped front fender create classic "endure" styling for off-road, endurance racing. Jeremy Cupp of LC Fabrications labored over this highly customized "Dirtster" for close to three years before finally declaring his stubborn streak of meticulous perfectionism to be satisfied.

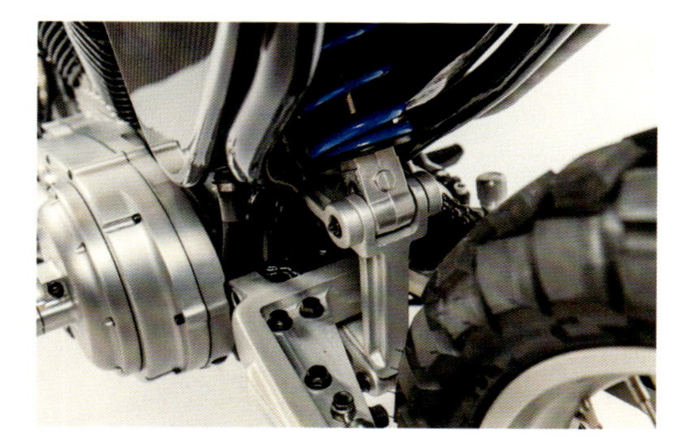

THE AMERICAN

1200CC (2011)

DEUS EX MACHINA

UNITED STATES

The first custom cycle ever produced in the United States by Deus Ex Machina, literally meaning, "god from a machine," was purchased by the Haas Moto Museum from actor Ryan Reynolds. It's a street legal café racer based on the chassis of the famous American dirt track racer "C&J Low Boy" and featuring a modified 1200cc Harley-Davidson Sportster V-twin with Edelbrock cylinder heads.

DEUS

DEUS

MANTA

650CC (2018)

JAY DONOVAN CANADA

Completed in less than three months in order to qualify for the 2018 "Motorcycles as Art" exhibit at Sturgis, the Manta features exquisite metalwork and remarkable detailing, down to each chromed acorn nut and domed bolt head, all in the classic Italian coach-building style so admired by designer and builder Jay Donovan.

YAMAHA

STINGRAY

(2019)

JAY DONOVAN

CANADA

A worthy successor to his highly acclaimed "Manta" custom cycle introduced at the 2018 Handbuilt Show, the Stingray was built by wunderkind Jay Donovan, who pushed the edge of traditional coach-building skills into the realm of electric motorcycles with this exemplar of innovative design and meticulous fabrication that has become his trademark.

AMADEUS

(2021)

JAY DONOVAN

CANADA

Named by the museum's founder after Wolfgang Amadeus Mozart to reflect the precocious genius of its twenty-something builder, Amadeus incorporates a symphony of form and function that completes the nonpareil trilogy by Jay Donovan. The addition of Amadeus follows the installation of its brethren Manta and Stingray, all three of which now grace the halls of the Haas Moto Museum. The dramatic curves of the hand-formed bodywork were designed, first on paper and then with metal, to achieve flow and balance that would embrace the elegant shape of its 1963 BMW R60/2 engine. The cycle's geometry simultaneously became both the blueprint and the constraints within which to form its stainless steel, trellis-style frame in a tortuous yet triumphant effort worthy of being the capstone of the Donovan trilogy.

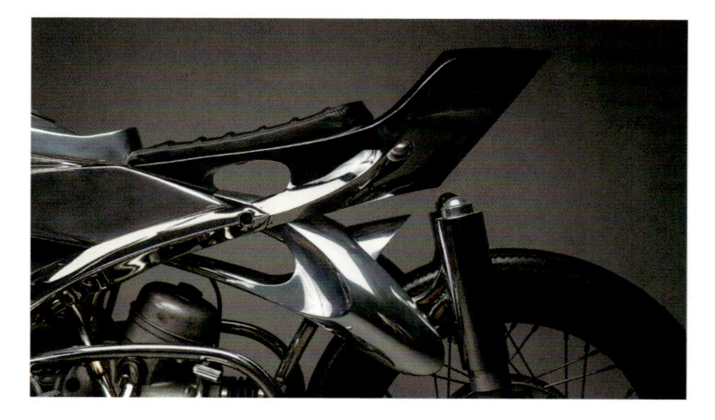

EARLE DUCATI 900 STREET TRACKER #2

900CC (2015) ALEX EARLE UNITED STATES

A designer for Porsche, Volkswagen, and Ducati, Alex Earle produced only two copies of this street tracker, following the rigorous auto design steps, from scale model to 3D scan to mold production to reality; the result was a 347-pound gem with 33-inch elevated seat height, flat track proportions, full carbon fiber monocoque design, and extended handlebars powered by a Ducati 900cc engine.

ECOSSE HERETIC

1753CC (2004)

DONALD ATCHISON OF ECOSSE

UNITED STATES

Fulfilling the company's promise of "first produced, last sold," the original Ecosse Heretic #01 of only 100 ever built, with just 147 miles on its odometer, finally sees the light of public display, living up to *Cycle World's* apt billing as a "blend of bare-fisted American muscle, Italian-inspired chassis, Swedish suspension, and thug-like riding position."

2029

(2019)

FULLER MOTO

UNITED STATES

Imagine what Georges Roy, designer of the iconic 1929 Majestic, would have built a hundred years later in 2029. That's exactly what Bryan Fuller and Bobby Haas imagined, and that's precisely what Fuller created in this masterful tour de force, using a combination of old-world craftsmanship with cutting-edge electric firepower and 3D printing for a host of the components.

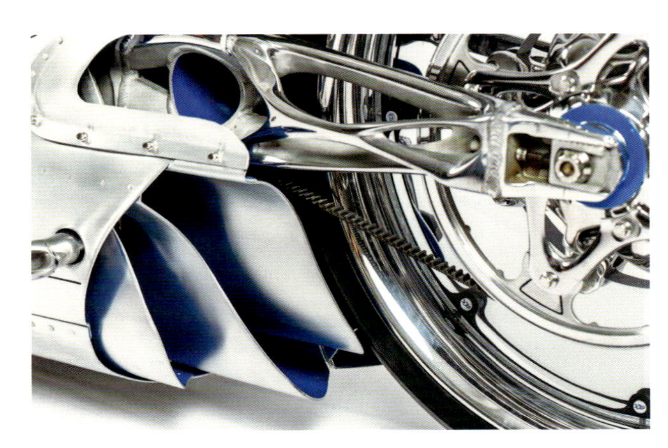

SHOGUN

600CC (2016)

FULLER MOTO

UNITED STATES

A trio of top-flight artisans devoted eight years of labor to the birthing of Shogun, before a fourth, a Samurai swordsmith, joined the crowd, knitting the threaded handgrips; each part of this masterpiece moved along a relay from moto fabrication by Bryan Fuller to the sketching of designs with a Sharpie pen by graffiti artist Totem to metal engraving by Tay Herrera and then back to Fuller for assembly.

ARP
ARP

Indian

CHIEF AMBASSADOR

1000CC (2018)

FULLER MOTO

UNITED STATES

Rivaling the eight-year marathon during which master designer/builder Bryan Fuller fashioned his legendary cycle "Shogun," Fuller devoted more than a handful of years to the one-off marriage of a 1969 Moto Guzzi Ambassador drivetrain with Indian fenders, tank, and frame from three different decades. Fuller cloaked his masterpiece in a bronze patina and fittingly anointed it with the moniker "Chief Ambassador" to pay homage to its dual American Indian and Italian Guzzi lineage.

MEDUSA

(2019)

SOFI AND GEORGE TSINGOS

UNITED STATES

The father-daughter team of George and Sofi Tsingos dreamt for years of this collaboration, and a commission from the Haas Moto Museum enabled the two to complete this blend of artful styling and engineering excellence, with intricate tubing to form its single-sided swing arm, Tickle-style front drum brake, and a rash of hand-formed features to complement its Yamaha 650 parallel twin.

THE ROCKET

2508CC (2016)

BOBBY HAAS AND STROKERS DALLAS

UNITED STATES

Unmistakable fingerprints of Art Deco design infuse all 125 inches along the arched frame of this 2508cc juggernaut, from its trio of bespoke grill-work to its 1938 Chevrolet headlamp.

THE KING

1638CC (2017)

BOBBY HAAS AND STROKERS DALLAS

UNITED STATES

Originally conceived as a custom café racer, a bundle of design and fabrication decisions radically converted this cycle into an elegant King of the Road that defies any attempt at being pigeonholed.

PREDATOR

1525CC (2015)

BOBBY HAAS AND STROKERS DALLAS

UNITED STATES

Hand-painted images of the four major predators of Africa adorn this chopper, uniting Haas's career as a *National Geographic* photographer with his exploits as a designer of custom motorcycles.

SALT SHAKER

1650CC (2019)

HAZAN-HAAS RACING

UNITED STATES

Conceived by Max Hazan and Bobby Haas, this "from scratch" speed merchant, built and piloted by Hazan around a Motus V4 pushrod engine, challenged the Bonneville Salt Flats in September 2019 and sprinted away with a new world record, topping 207mph. In the process, the Salt Shaker added a fresh chapter to the legendary saga of Max Hazan, nonpareil custom builder.

FIRST BIKE

389CC (2011)

HAZAN MOTORWORKS

UNITED STATES

Conceived as no more than an experiment, this creation launched the meteoric rise of Max Hazan to the very highest echelon of custom builders; it has known only two homes, Max's living room and the Haas Moto Museum.

SUPERCHARGED IRONHEAD

900CC (2015)

HAZAN MOTORWORKS

UNITED STATES

Powered by a supercharged 900cc Harley-Davidson Ironhead Sportster, this motorized piece of jewelry by Max Hazan was named "Custom Motorcycle of the Year" for 2014 by Pipeburn.com.

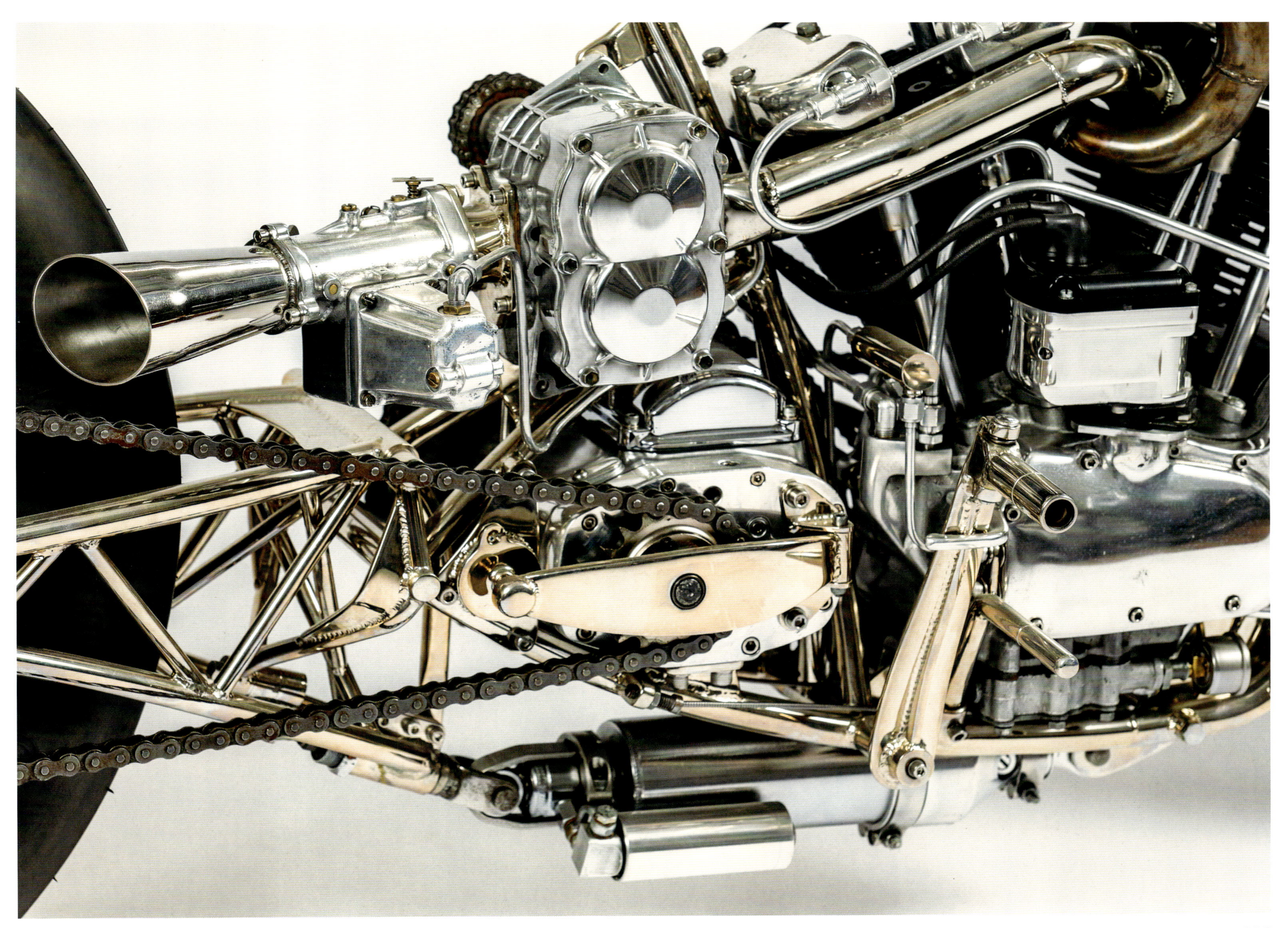

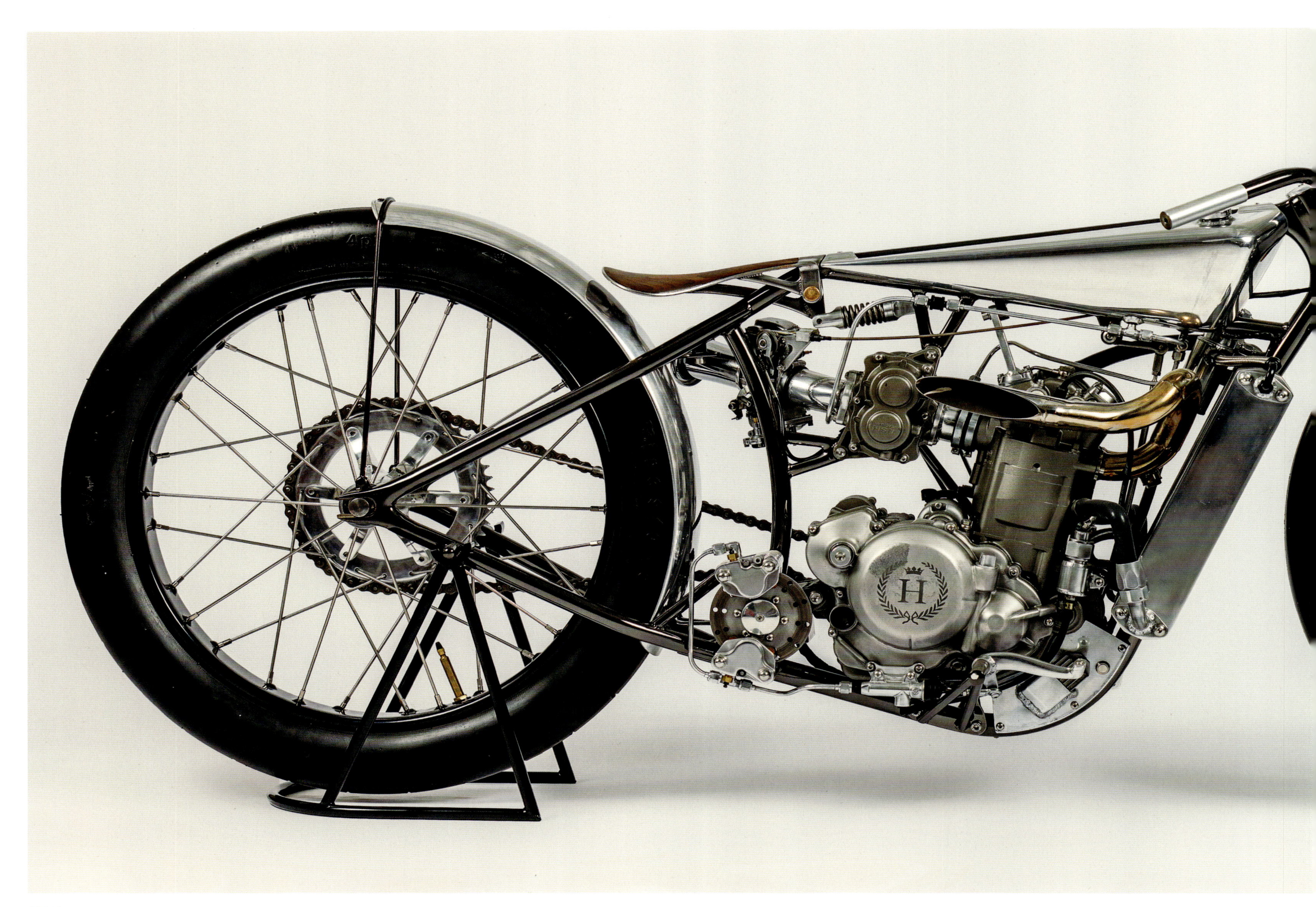

SUPERCHARGED KTM

520CC (2017)

HAZAN MOTORWORKS

UNITED STATES

On its way from Max Hazan's shop to the Haas Moto Museum, this newborn dazzled the crowds at Sturgis and then stopped over in Las Vegas just long enough to capture the coveted 2017 Artistry in Iron crown.

TWIN MUSKET

(2015)

HAZAN MOTORWORKS

UNITED STATES

The Hazan legend grows as this uniquely configured 1000cc Twin Musket earns Max Hazan an unprecedented third consecutive "Custom Motorcycle of the Year" honor for 2015 from Pipeburn.com.

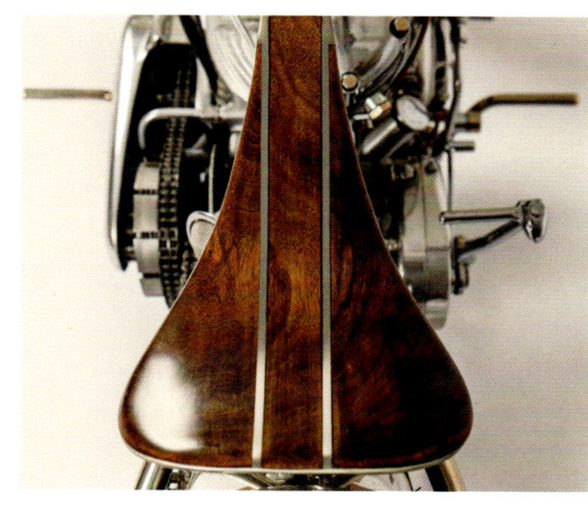

THE MUSKET
COLUMBUS OH

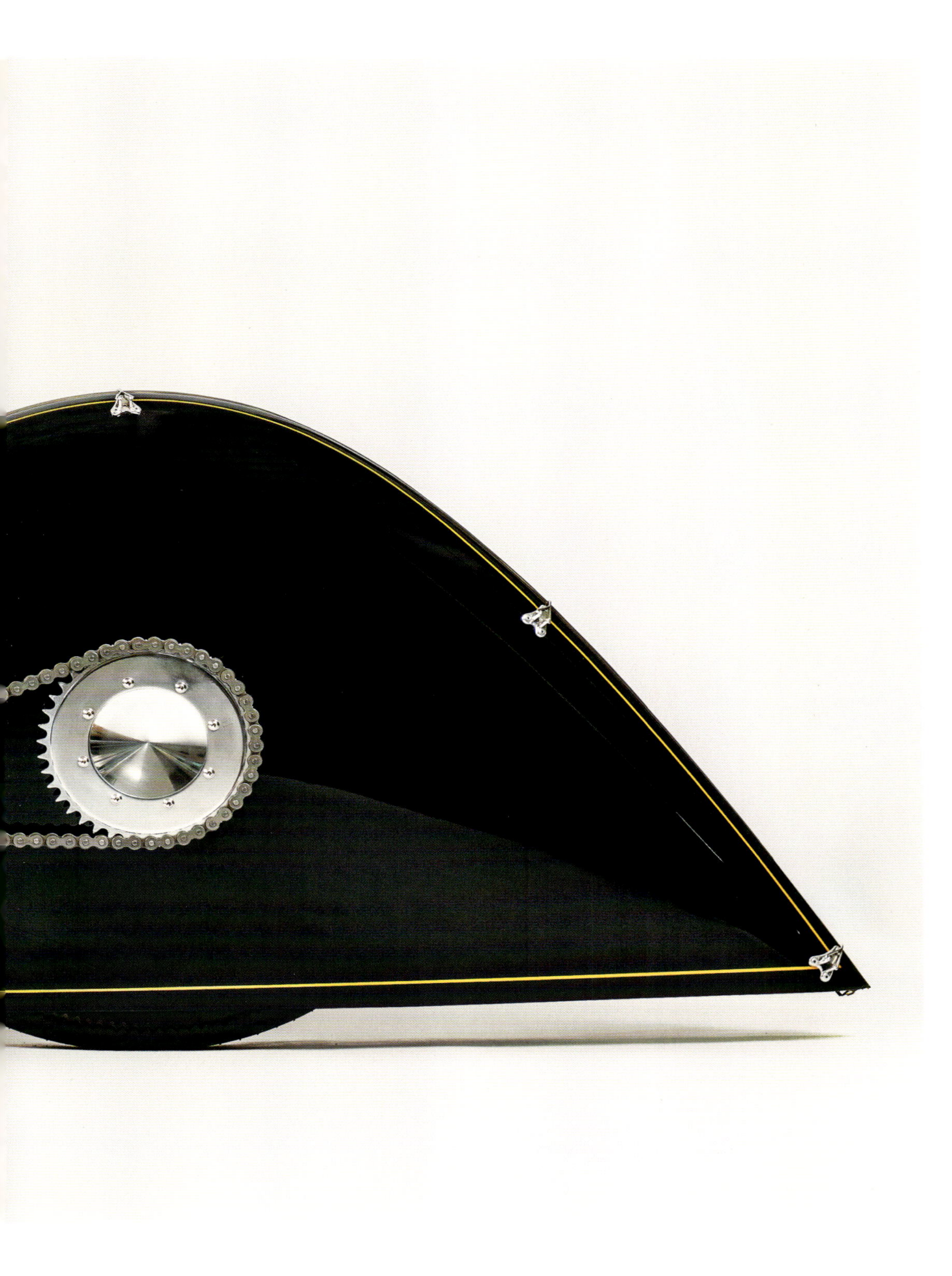

BLACK KNIGHT

500CC (2016)

HAZAN MOTORWORKS

UNITED STATES

When the wunderkind designer/builder Max Hazan, winner of three consecutive Motorcycle of the Year awards and the 2017 Artistry in Iron competition, refers to the Black Knight as his finest creation ever, it merits a close look.

GUTE GEIST

(2020)

DIRK OEHLERKING

GERMANY

Conceived as the patriarch of the Phantom family of custom creations by master craftsman Dirk Oehlerking, "Gute Geist," or the "Good Ghost," like the White Phantom and Black Phantom before it, is powered by a BMW boxer-style engine, in this case an R100RS Type 247, vintage 1980. Gute Geist features elongated, hand-formed bodywork that is the stunning hallmark of the Phantom series, leaving no doubt that its lineage traces its roots to the same workshop in Gelsenkirchen, Germany, that fashioned the White Phantom in 2016 and the Black Phantom in 2018.

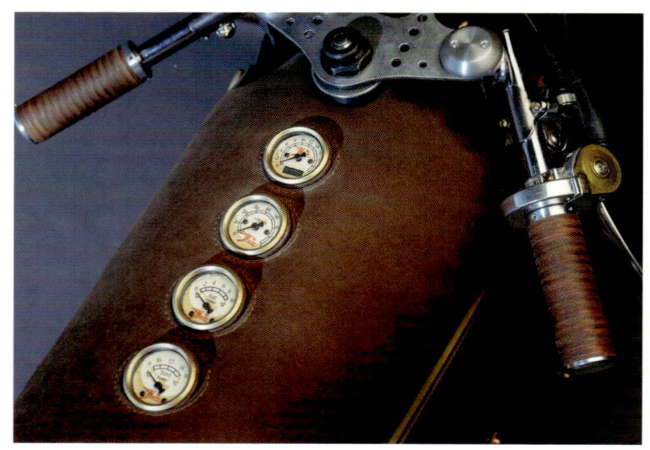

CHOKE
NGK

Kingston

HOMMAGE

(2022)

DIRK OEHLERKING

GERMANY

"A Kingston family; related but each member individualistic." The final custom motorcycle to be commissioned by Bobby Haas. He challenged master craftsman Dirk Oehlerking of Kingston Customs to build "the longest custom BMW boxer of all time," that would complement the three other Kingston gems in the Haas Moto Museum. Originally known as "Eleganza" due to its Art Deco-era streamliner design, the project took a swift turn after Bobby's sudden passing in September of 2021, prompting the name of the build to change to "Hommage" as a tribute to the founder of the museum. The custom aluminum bodywork swaddles a 1977 BMW R100, complemented by other distinct parts on this one-of-a-kind, 12-foot 6-inch build, such as the taillight from a BMW 700, the mirror from a Porsche 356 Speedster, and a custom BMW-style front grill.

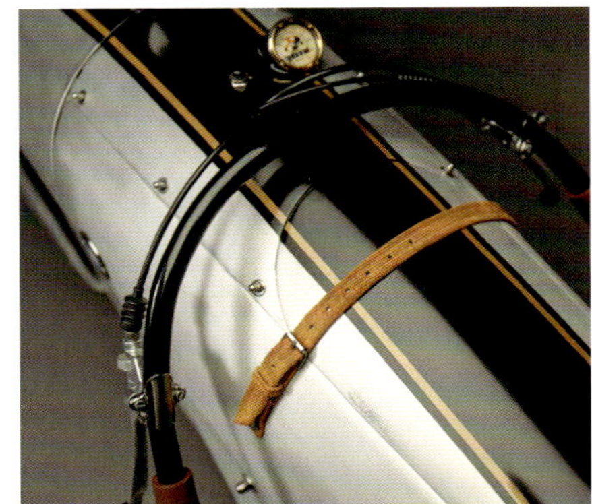

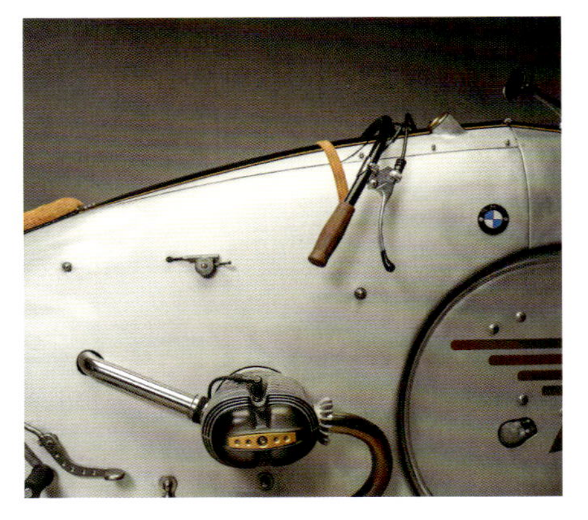

WHITE PHANTOM

797CC (2016)

DIRK OEHLERKING GERMANY

Fresh off an engagement as one of the honored cycles in the "Custom Revolution" exhibit at the Petersen Automotive Museum, the White Phantom by Dirk Oehlerking of Germany headed to Texas to grace the Haas Moto Museum as a permanent resident of its Custom Shop. The beauty and refinement of its passel of customized features envelop a rebuilt 1986 BMW R80 boxer-style twin, complemented by a swing arm that encloses its shaft drive. The pièce de résistance of this dazzling creation must certainly by its exquisite bodywork, highlighted by an upper section housing the gas tank and studded with a row of circular gauges that swings open to reveal the turbo-charged power station below.

BMW
Kingston

BLACK PHANTOM

745CC (2018)

DIRK OEHLERKING

GERMANY

On the heels of the worldwide acclaim of his White Phantom, designer/builder Dirk Oehlerking was challenged to create a sibling worthy of the fame that regularly emanates from his Kingston Custom shop in Gelsenkirchen, Germany. And the Black Phantom proved to be a most worthy successor. Exquisite bodywork, this time in ebony to complement the ivory of the White Phantom, encases its 1979 BMW R80 boxer twin. Every component of the Black Phantom was sourced by Dirk from parts dating from 1951 to 1979. In a twist of delicious irony, the White and Black Phantoms were never before under the same roof until the siblings were united as permanent exhibits in the Haas Moto Museum.

BMW

CHERRY BLOSSOM

750CC (2013)

MITSUHIRO (KIYO) KIYONAGA

UNITED STATES

The first in a trilogy of racers built around Honda CB750 four-cylinder engines, Cherry Blossom launched Kiyo on an improbable seven-year mission to design, fabricate, and race three custom-built bikes, each time adding another Honda CB750 engine to power his creation.

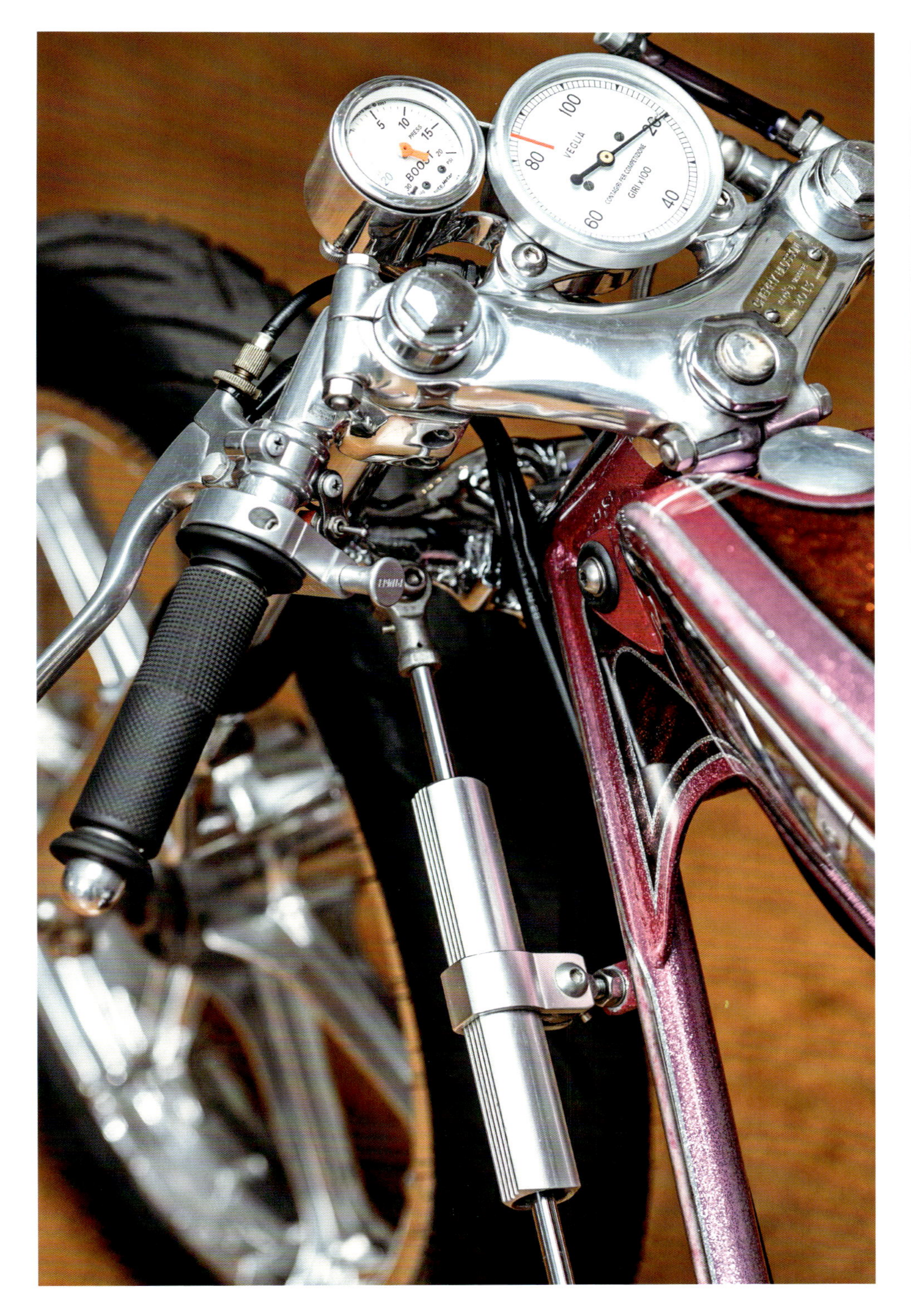
BOOST
VEGLIA
GIRI x 100

ATP
ATP

GEKKO

1620CC (2016)

MITSUHIRO (KIYO) KIYONAGA

UNITED STATES

The second in the Kiyo trilogy of racers featuring Honda CB750 engines, Gekko's power station consists of two Honda CB750 engines, each overbored to 810cc for total displacement of 1620cc. This stunning sibling of Cherry Blossom sprinted across El Mirage Dry Lake at 173mph and captured first place for Invited Builders and Best Japanese Bike at the Born Free 8 Custom Motorcycle Show.

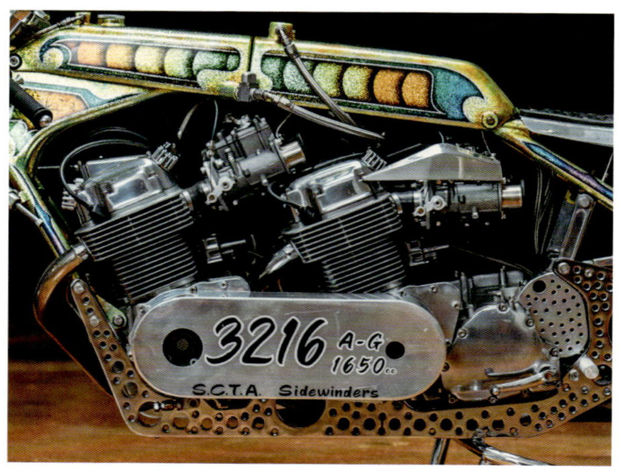

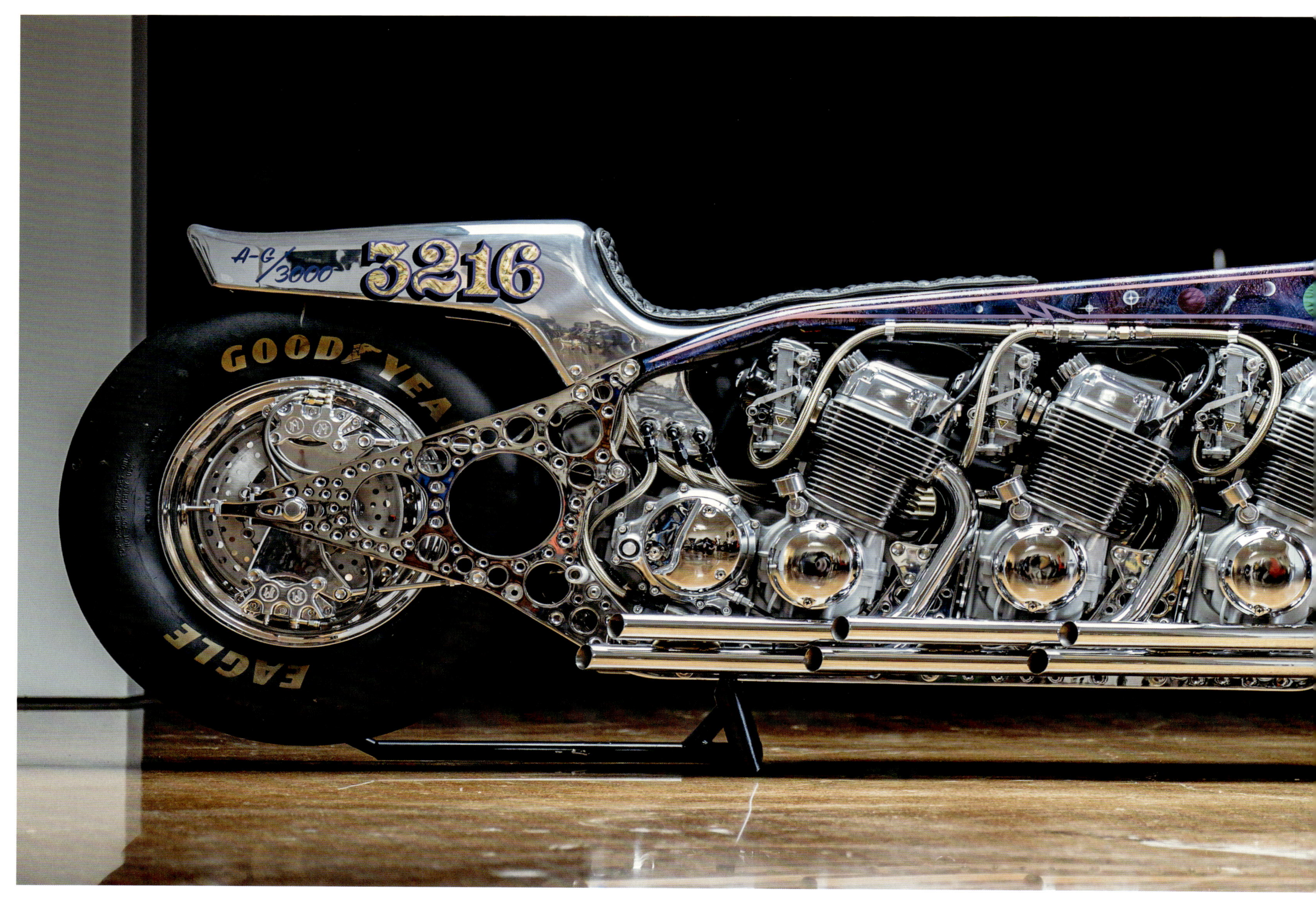
A-G/3000
3216
EAGLE

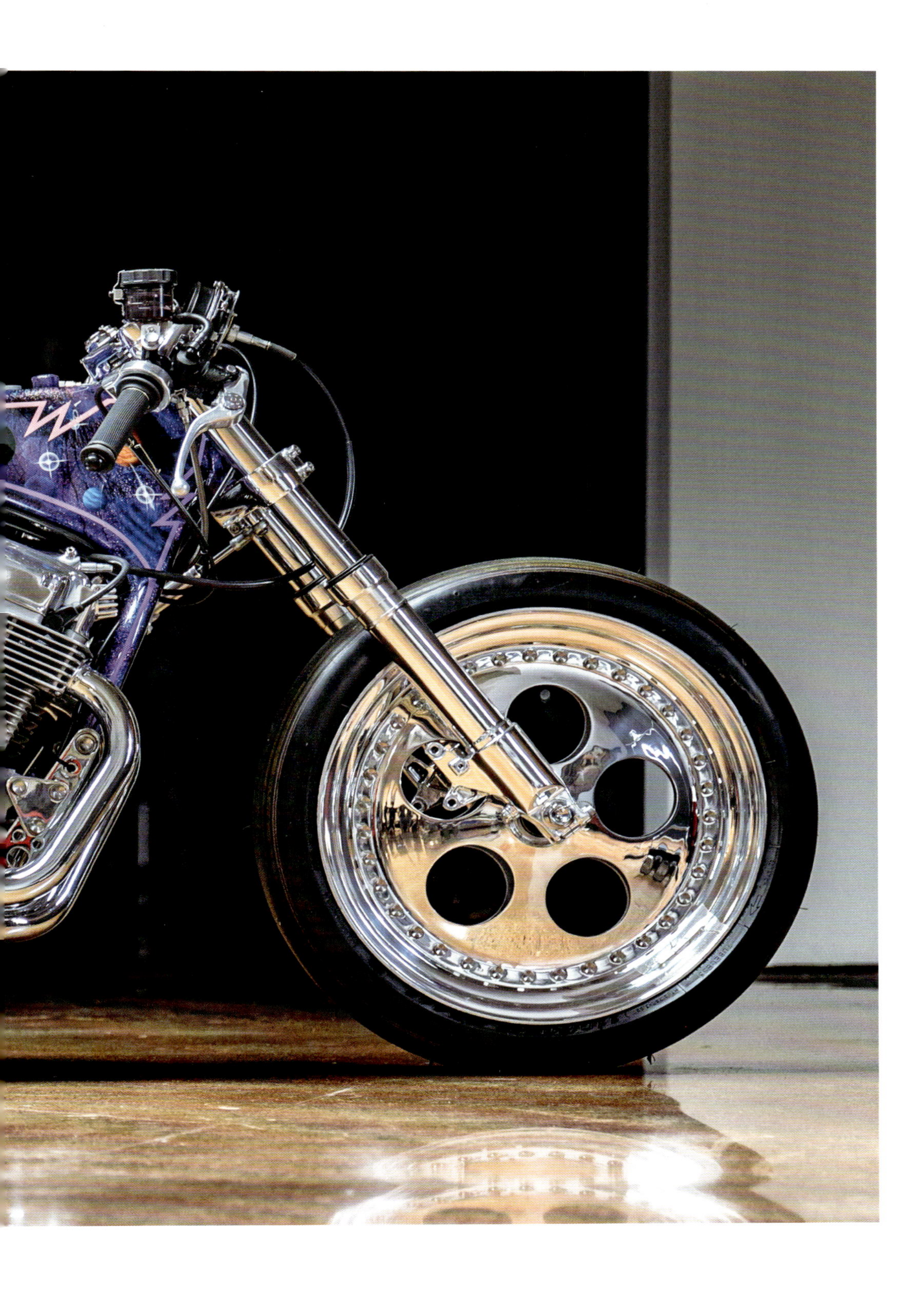

GALAXY

2508CC (2020)

MITSUHIRO (KIYO) KIYONAGA

UNITED STATES

Beset by a host of logistical and engineering challenges throughout the pandemic of 2020, master craftsman Matsuhiro "Kiyo" Kiyonaga struggled to fashion this three-engine, twelve-cylinder racer with an overbored displacement of over 2500cc. What emerged from that ordeal was this landmark creation, capping the immediately famous Kiyo trilogy of Honda CB750 racers and fulfilling his lifetime dream of progressing from one engine to three in a series of stunning land speed racers.

CUSTOM KAWASAKI W1R

624CC (2016)

MICHAEL LAFOUNTAIN

UNITED STATES

A seven-year odyssey by Michael LaFountain of Raccia Motorcycles finally reacheds its ultimate destination in the Haas Moto Museum. Inspired by a single grainy photograph of a Kawasaki W1R racer, itself based on the 1966 W1 street bike, LaFountain commited himself to recreating the obscure racer, utilizing only original Kawasaki parts from different bikes and different eras. Dogged by frustration, endless stress, and unforeseen setbacks, LaFountain was often tempted to throw in the wrench and abandon the challenge of recreating this factory racer, Kawasaki's own effort from decades earlier to duplicate the Matchless G45. But his reservoir of passion, creativity, and skilled workmanship eventually carried the day, as LaFountain succeeded in creating a brilliant work of moto artisanship that bears none of the scars of his Frankenstein-like process of "lashing-up" the legendary racer.

KAWASAKI

Indian

THE FLYING SQUIRREL

616CC (2017)

JON MACDOWELL

UNITED STATES

Utilizing a "basket case" engine and transmission from a 1941 Indian 741 Scout, designer/builder Jon MacDowell fashioned this custom bike with a combination of stainless steel, aluminum, and brass. The dropped handlebars and exquisite hand-formed aluminum tank instill this customized bobber with a race-style design that hearkens back to an earlier era and yet bears the unmistakable signs of elite modern craftsmanship.

RONDINE

580CC (2013)

MEDAZA CYCLES

IRELAND

Designed and fabricated by a small shop near Cork, Ireland, Rondine stunned the custom motorcycle world (and even its creators, Don Cronin and Michael O'Shea) when it captured the 2013 AMD World Championship in Essen, Germany. The Medaza founders, heavily influenced by their sculptural expertise, create only a few cycles that embody their raw passion for rolling art. Rondine's power station, a 1974 Moto Guzzi Nuovo Falcone flat single, overbored to expand its displacement from 500cc to 580cc, is encased in a customized frame with bodywork of hand-formed aluminum and a cornucopia of one-off components.

Medaza

PORTERFIELD

1442CC (2018)

MOTORCYCLE MISSIONS

UNITED STATES

Under the guiding hand of Krystal Hess and Motorcycle Missions, the non-profit that Krystal founded to aid in the recovery of veterans and first responders suffering from PTSD, a group of eight veterans converted a 1974 Harley-Davidson Shovel into this sleek board track racer with vintage flair. The resulting masterpiece of elegant design, with minimalist frame and retro-style handlebars, won the highly coveted first place award in the freestyle category at the 2018 J&P Cycles Ultimate Builder Custom Bike Show. The number "22" on the rear wedge of the frame is a sobering reminder that on average, twenty-two military veterans succumb each day to paying the ultimate price for their service by taking their own life.

MOTORCYCLE
MISSIONS
MADE IN USA

TRIUMPH

BONNIE

(2018)

ROB CHAPPELL OF ORIGIN8OR

CANADA

In a converted garage-workshop on the outskirts of Toronto, designer-builder Rob Chappell of Origin8or Custom Cycles deliberately set the bar higher than before against an impossible deadline to create this highly stylized version of a Triumph Bonneville Board Tracker, featuring a 1969 650cc parallel twin, sleek body-work, and a bevy of handcrafted parts and assemblies.

CLYDE

(2017)

ROB CHAPPELL OF ORIGIN8OR

CANADA

Just a few feet away from the custom board tracker "Bonnie" in Rob Chappell's workshop stood this customized version of the Honda CB750 inline-four racer, itself the winner of the Café Racer Class at the 2018 International Motorcycle Show in Toronto. Unable to resist the temptation to keep the two customs side-by-side, the Haas Moto Museum decided to head back to Dallas with both in tow, attaching the name "Clyde" to the Honda racer so that "Bonnie" and "Clyde" would continue to keep each other company.

CR
KEIHIN

13

ALUMINATI

125CC (2016)

PARKER BROTHERS CONCEPTS

UNITED STATES

Dream-meisters Marc and Shanon Parker designed and built the Aluminati in the incredible span of only ten days, just in time to debut at the 2016 Daytona Bike Week and capture the Judges Choice Award.

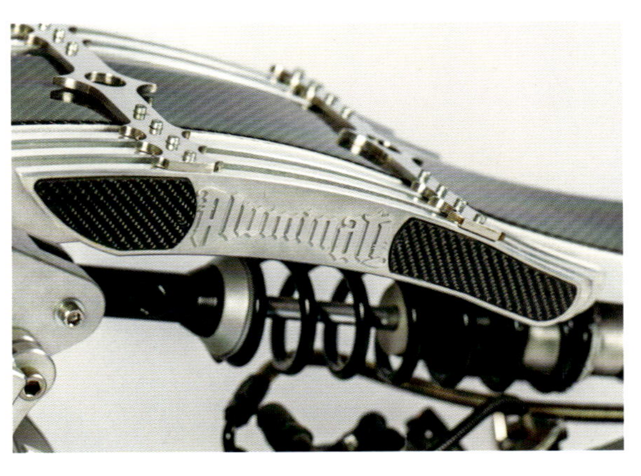

NEUTRON

(2017)

PARKER BROTHERS CONCEPTS

UNITED STATES

This customized electric-powered Neutron features a mesmerizing pair of hubless wheels supporting its futuristic profile; this model dispenses with the niceties of "street legal" requirements to offer a raw, unvarnished, but fully operable glimpse into the future.

pbc

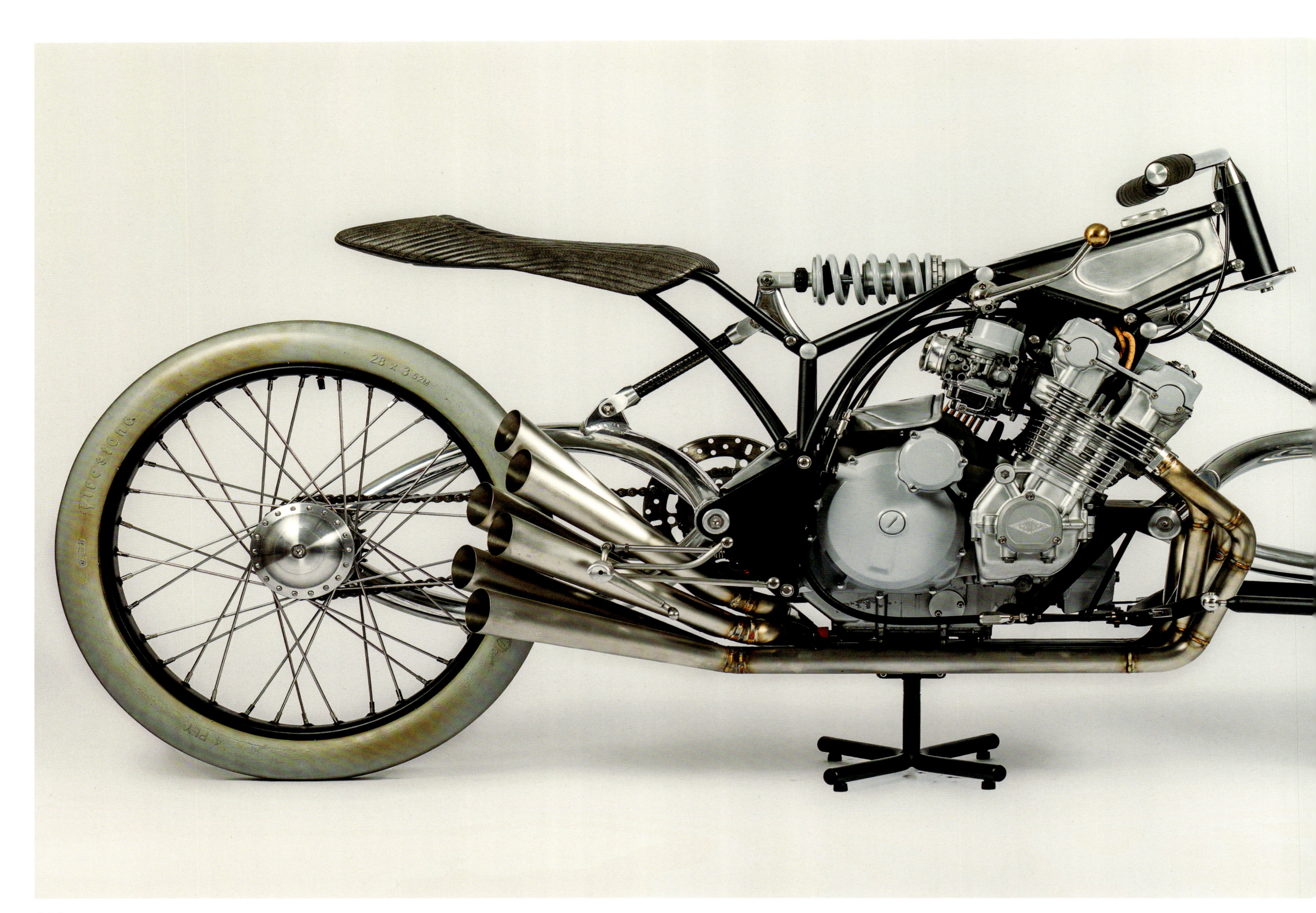
Firestone
28 x 3

THE SIX

1047CC (2018)

REVIVAL CYCLES

UNITED STATES

Originally conceived by Alan Stulberg and Bobby Haas and painstakingly translated into a masterwork by the expert craftsmen and engineers of Revival Cycles, "The Six" features avant-garde styling and hub-centric steering that envelop its 1047cc Honda CBX inline six-cylinder engine in a metallic magnum opus.

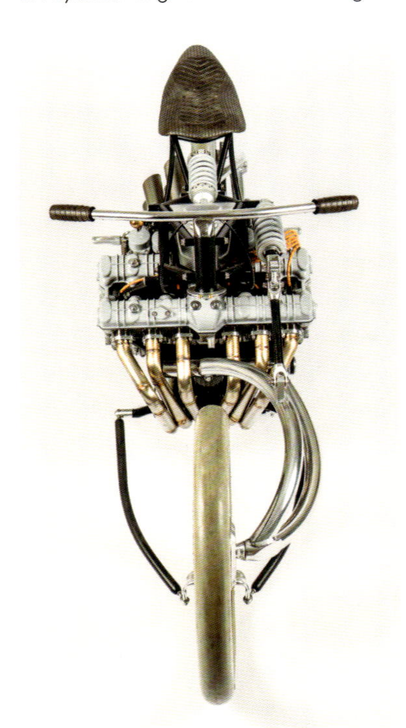

THE RACER

(2015)

REVIVAL CYCLES

UNITED STATES

Inspired by the 1928 Henne BMW Landspeeder, with a distinct hint of influence by the one-and-only 1934 BMW R7 concept cycle, this work caused the uber-authoritative website Bike EXIF to gush that The Racer had "smashed the mold [of custom builds] to smithereens."

REVIVAL

RK FIGHTER

1200CC (2011)

RK CONCEPTS

UNITED STATES

Armed with a Buell 1200cc engine and surrounded by "bullet-holed" metal, the RK Fighter appears to be designed more for the rugged challenges of war than for the flowing ribbon of street gliding.

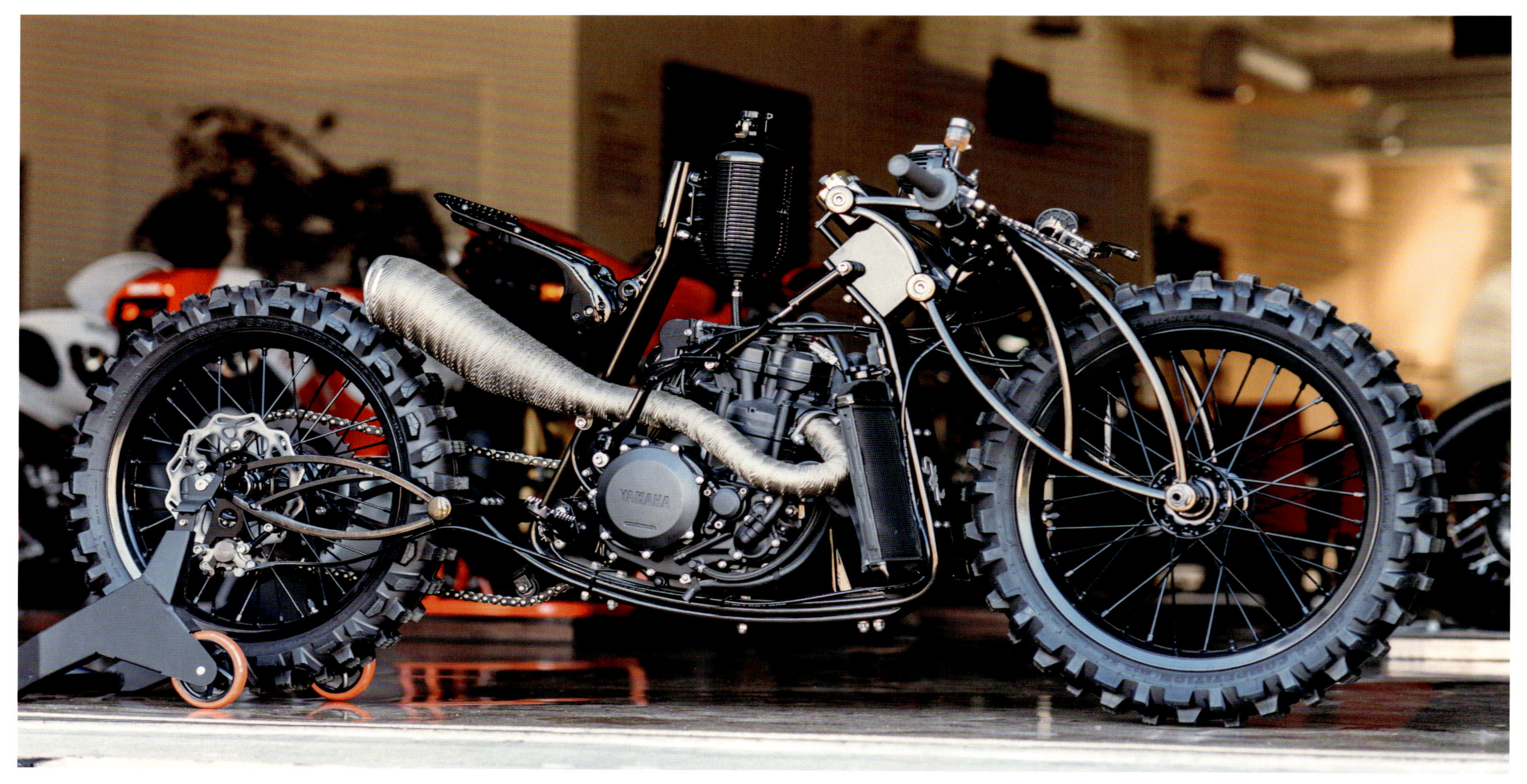

RK SPRING

(2015)

RK CONCEPTS

UNITED STATES

This highly innovative cycle, distinguished by a vertical gas tank and an undergirding of leaf springs that double as the cycle's frame and its suspension, is one of three radical creations that designer Rafik Kaissi has entrusted to the Haas Moto Museum.

VIS-À-VIS

449CC (2015)

RK CONCEPTS

UNITED STATES

The fertile mind and talented hands of Rafik Kaissi transform into working reality mirror images of the front and back halves of this cycle, with a suspended third wheel cradling the engine.

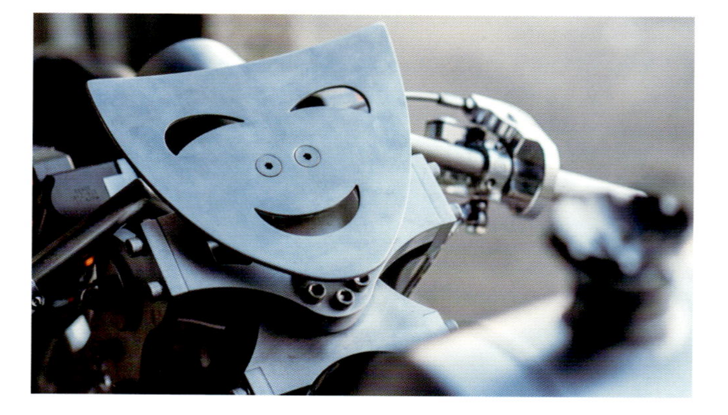

THE AMBASSADOR

750CC (2016)

CRAIG "RODSMITH" PRESSNELL

UNITED STATES

Its tires had barely touched the pavement at its first major show, the 2017 Handbuilt Show in Austin, when this turbocharged Moto Guzzi, with its arresting, handcrafted aluminum fairing, was snatched up by the Haas Moto Museum.

CORPS LÉGER

150CC (2018)

CRAIG "RODSMITH" PRESSNELL

UNITED STATES

After the Corps Léger was conceived and crudely sketched at a design session between Bobby Haas and Craig "Rodsmith" Pressnell in mid-2017, Pressnell successfully took on the challenge of demonstrating how moto functionality and metallic beauty could be merged in the narrowest space possible.

BSA

THE KILLER

(2019)

CRAIG "RODSMITH" PRESSNELL

UNITED STATES

Inspired by a few grainy photographs of a concoction from the 1930s by a group of German engineers, Bobby Haas cajoled Craig "Rodsmith" Pressnell into taking on a challenge that seemed sheer folly at the time: build from scratch an Art Deco masterpiece of design and engineering with a three-cylinder engine somehow embedded in the front wheel—and manage to accomplish this without any CNC machinery on a seventy-year-old manual lathe and mill—and then wrap this engineering marvel in the exquisite handcrafted aluminum bodywork that only Rodsmith can fashion. And the Aussie master craftsman responded by producing a motorcycle for the ages.

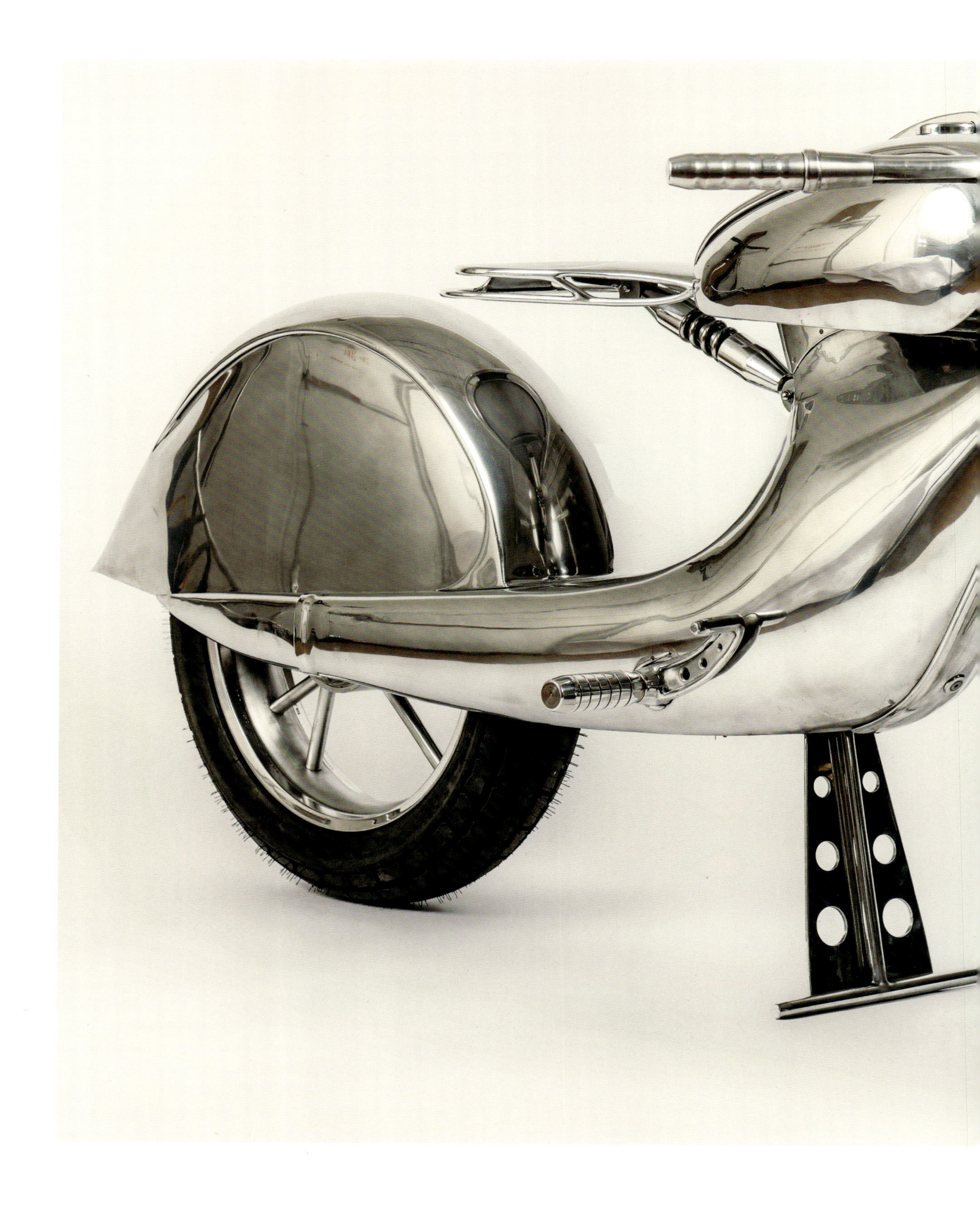

RCK

(2013)

CRAIG "RODSMITH" PRESSNELL

UNITED STATES

Built around a KZ750 Kawasaki parallel twin, RCK helped define the signature style of custom moto genius Craig "Rodsmith" Pressnell, with strong elements of bare metal and polished hand-formed aluminum, accented by the distinctive styling of the curved frame backbone and fuel tank. The nickel plating was applied by Rodsmith himself and not outsourced to a plating specialist, and the split-front fairing is reminiscent of the windshields on World War II fighter planes. The total effect is a custom build that manages to straddle the twin appeal of aggressive styling and elegant craftsmanship.

NGK

MISTER FAHRENHEIT

(2020)

CRAIG "RODSMITH" PRESSNELL AND BOBBY HAAS

UNITED STATES

Originally conceived by Craig "Rodsmith" Pressnell and Bobby Haas as a sidecar outfit that would cruise along the highway when it was not on display in the Haas Moto Museum, this custom creation gradually morphed into a serious racing contender whose sole mission in life would be to chase a land speed record. Its deep-throated Moto Guzzi power station was adjusted to coax maximum velocity out of a displacement of just under 1000cc. The aluminum bodywork was fashioned in the classic Rodsmith style, and the companion sidecar was reduced to this minimalist configuration. Poised to take on the challenges of America's speedways in 2021, Mister Fahrenheit is yearning to defy the laws of gravity in its quest for racing glory.

TOCHTLI

600CC (2021)

CRISTIAN SOSA

UNITED STATES

Tochtli (Aztec for "rabbit") is the creative culmination of hundreds of hours at the hands of master metal craftsman Cristian Sosa. An extremely rare ninety-seven-year-old Douglas horizontally opposed twin-cylinder race engine has been married to a 1958 three-speed Triumph transmission. The 9hp powertrain is embedded in a "from scratch" flat plate steel frame, designed to conduct airflow as its RAM induction system feeds the AMR 500 supercharger. A pair of identical drum brake wheels perfectly balance the bike's styling, while a left-hand throttle complements the right-hand jockey shift. To keep the handlebars as lean as possible, the floorboards were modeled to include foot pedal rear braking and clutch controls.

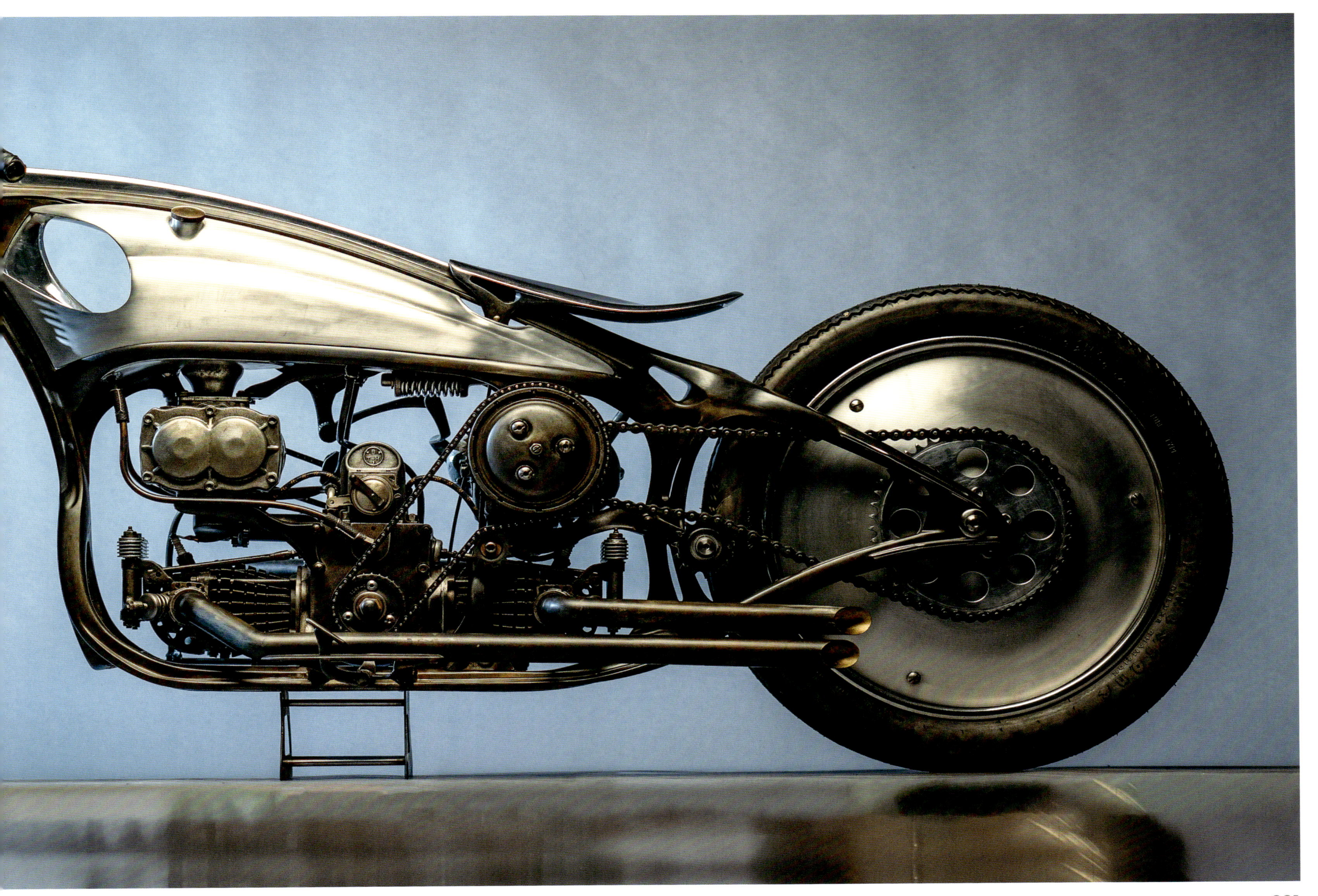

SPACE TRAVELER

1200CC (2017)

CRISTIAN SOSA

UNITED STATES

Master of precision metalwork, Cristian Sosa applied his renowned talents to defying the passage of time by enveloping a 1946 Knucklehead V-twin in a revolutionary frame with ovalized tubing, suspending the front end with a homemade leaf spring fork, and finishing the skin of this award-winning creature with a brushed metal patina that imprinted the Space Traveler with his signature style.

MSM V-TWIN SPEEDWAY BIKE

1442CC (2018)

PATRICK TILBURY

UNITED STATES

This customized V-twin speedway bike boasts a unique tale that hearkens back half a century. The major components of the engine are from "the one-and-only" Meierson Sprint Motor MSM 1000cc V-twin produced in 1967 by the Australian father-son team Clarrie and Allan Meiers for sidecar speedway competition. In its first season, the V-twin placed first in all fifteen of its events. Shortly thereafter, the famous engine disappeared, only to resurface decades later in the biker shop of American television personality Jesse James, where it was spotted and purchased by Patrick Tilbury. The resourceful Tilbury tracked down Allan Meiers, by now in his seventies, who verified that the engine was indeed "the one-and-only" MSM V-twin. With advice from the surviving Meiers, Tilbury brought the MSM engine back to life as the power station for this dazzling speedway racer.

Royal-T
M M S
AUSTRALIA
LECTRON

BEDEVELED

(2019)

WALT SIEGL MOTORCYCLES

UNITED STATES

Commissioned by the Haas Moto Museum to push the limits of his acclaimed racer-style craftsmanship, Walt Siegl utilized a Ducati bevel engine from decades past to create the type of racer that Ducati lovers would have yearned to see on the track in the 1970s and '80s, designing and building a host of modernized features that flesh out this trellis-framed masterpiece, all the way from its bespoke fairing to its coiled rear suspension.

Agip
R 4810

WSM PACT PROTOTYPE

(2019)

WALT SIEGL MOTORCYCLES

UNITED STATES

Master designer-builder of a world-renowned progression of limited-production, internal combustion cycles in the European tradition, Walt Siegl teamed with Mike Mayberry of Ronin Motorworks fame to produce this prototype electric cycle. Outfitted with an Alta Redshift drivetrain and liquid-cooled, battery-powered motor, the PACT will launch a limited production line of e-cycles by Siegl that will carry his unmistakable fingerprint of a velocity-rich machine lacking only a rider.

RONTU

(2021)

WALT SIEGL MOTORCYCLES

UNITED STATES

Challenged by his own inner drive and by the Haas Moto Museum to craft from scratch a visually arresting and ultralight e-powered cycle with structural integrity that would withstand the rigors of speed and the demand for flexibility, Walt Siegl committed to this project with unprecedented devotion. Electric cycle technology enabled Siegl to experiment along the way, while utilizing a monocoque chassis design that required large surfaces to provide the structural stability needed for the cycle. Ultimately, Siegl was able to depart from the maxim that a motorcycle's design is defined by the contours of its bodywork, to build a cycle that is defined more by the lack of bodywork. Relying on a frame skeleton composed of thin-walled chrome moly tubing boxed with carbon fiber and a host of aluminum parts, Siegl's final product weighs in at a mind-boggling 100 pounds, almost defying the laws of gravity with its unimaginable lightness.

moto 1

MOTO 1

650CC (2020)

BOB STEINBUGLER

UNITED STATES

Spanning thirty years from conception to final completion, the MOTO 1 by Bob Steinbugler is powered by a 1988 Honda Hawk liquid-cooled, carbureted 650cc V-twin engine. After a twenty-eight-year hiatus, during which MOTO 1 was left virtually untouched, the Haas Moto Museum commissioned Steinbugler to complete his radical creation with a bevy of enhancements that lift MOTO 1 to even more dazzling design heights.

VYRUS 986 M2 TRICOLORE

600CC (2014)

VYRUS

ITALY AND UNITED STATES

This 323-pound speed demon, with rare hub-centric steering, lays claim to its position among other custom cycles in The Haas by virtue of the fact that its bespoke Vyrus parts are fabricated in Italy, married in the States to a Honda CBR600RR engine, and then finished with an Italian tricolore design unique to this one motorcycle. All Vyruses are individually customized for each purchaser. Of the approximately forty-three Vyrus 986 cycles produced to date, the majority were built for racing as qualified Grand Prix Moto2 competitors, while only about fifteen were built to be street legal, including this stunning and unique example.

SMOKING BLACK FISH

350CC (2015)

GEORGE WOODMAN

FRANCE

A remarkable hybrid of vintage provenance and modern artistic flair, this rugged Belgian FN Type M70B from 1929 boasts a proud heritage as part of "The Sahara" line of Belgian cycles celebrated for crossing the 3,000-mile Sahara Desert. In the hands of master craftsman George Woodman, the FN was enhanced with a new aluminum, wood, and leather seat unit, hand-formed wooden cowl housing an offset headlight, a wooden guard for the single's velocity stack, and other bespoke features, including the hand etching of a fish on its fishtail exhaust. Woodman has produced only a few custom cycles from his shop in Biarritz, France, and the Smoking Black Fish is the first one ever sold, after Woodman approached The Haas Museum and expressed his desire to have this creation installed in the museum.

FN

TYPE 9 SHOVEL

1524CC (2016)

ZERO ENGINEERING

JAPAN AND UNITED STATES

One of only twenty-four Type 9 Shovels scheduled for production in 2016 by Zero Engineering, designed in Japan with old-school aesthetics and built at Zero's Las Vegas production facility.

S&S
SUPER
STOCK
MADE IN U.S.A.
XNP00321X

力

THE SCULPTURE GALLERY

Throughout the entire museum is a diverse array of metal sculptures that echo the artistry of the motorcycles that stand nearby, ranging from a series of miniature "found metal" bikes painstakingly constructed piece by piece, to the life-size "Through the Wall" trio of motorcycles crashing through walls of concrete, glass, and plaster.

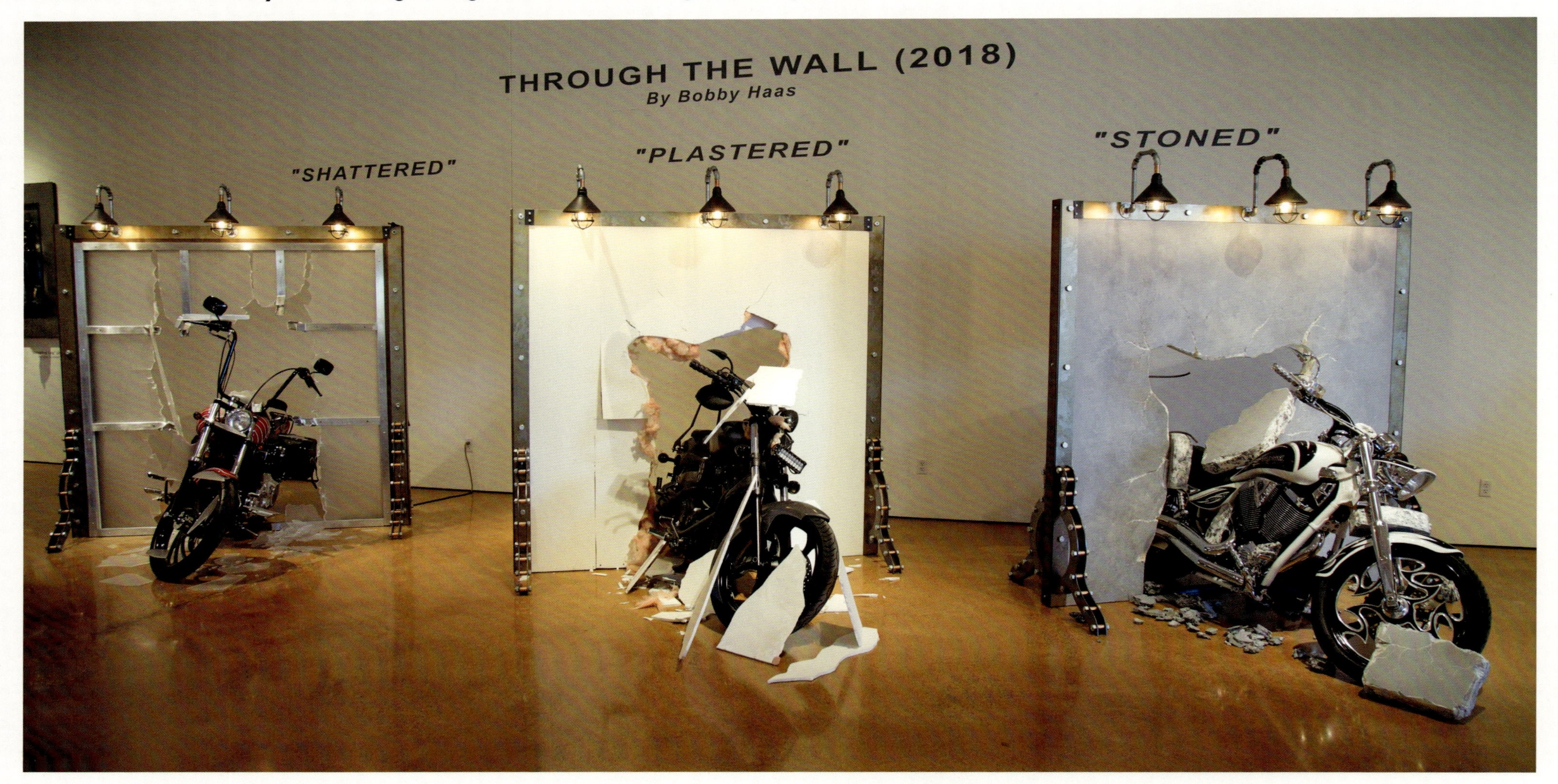

"THROUGH THE WALL" BY BOBBY HAAS

This three-dimensional art piece by Bobby Haas features a trio of motorcycles (from left to right: 1999 Harley-Davidson Wide Glide 1338cc; 2014 Yamaha XVS1300 Custom Cycle 1300cc; 2009 Victory Jackpot, Cory Ness Edition 1731cc) crashing through walls of glass, plaster, and concrete.

"SALVAGING HIDDEN TREASURES" BY MICHAEL ULMAN

It is often said that one man's trash is another man's treasure. No found-metal sculptor exemplifies this with more dazzling expertise than Michael Ulman, who learned this rare craft by welding scraps of metal into objects as a boy. Sourcing materials from junkyards, dumpsters, and trash heaps, Michael transforms detritus into hidden treasures. Rusty baby carriage wheels are reincarnated on a motorcycle, an upside-down shoe tree becomes a seat, a discarded vacuum cleaner finds fresh life as a sidecar, vintage ice tongs are repurposed as handlebars. The possibilities are limited only by the imagination, and Michael's imagination seems limitless. His talents were showcased in the blockbuster Australian film *"Mad Max: Fury Road"* featuring the infamous flame-throwing guitar designed and constructed by Michael. The results of his work are so lifelike that often our guests will ask, "Do they work?" And the answer is a resounding "Yes!"—if your definition of "working" is to inspire you to believe that anything is possible when genius is applied to scrap.

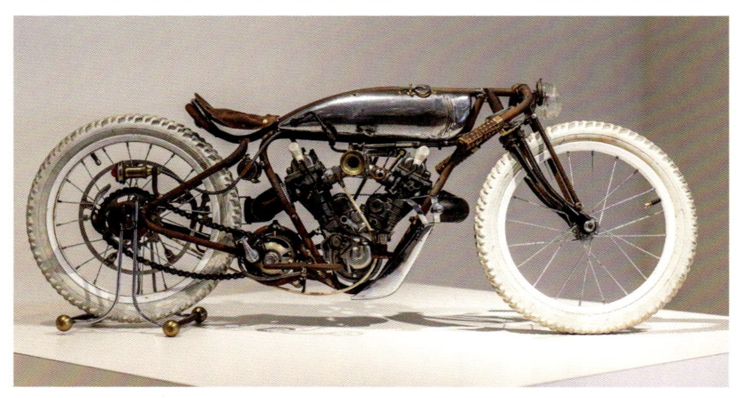

ENGINE SCULPTURES BY BOBBY HAAS

From left to right:

"Triple Threat" (2016)
1971 Kawasaki H1 Mach III (The Widowmaker)

"Greater Kudu Horns" (2016)
2013 Harley-Davidson V-Rod Revolution

"Double Cross" (2016)
2003 Harley-Davidson V-Rod 100th Anniversary Edition

"JACKET WALL"
BY BOBBY HAAS

Nine mannequins in unique helmets and Bobby's personal jackets, organized together in the fashion of a motorcycle gang cruising the open road, adorn the walls of the Custom Shop.

"CHAIN SCULPTURE"
BY JOHN FELDER

A Harley-Davidson Panhead, painstakingly fashioned over a handful of years by John Felder, a welder from Mississippi, using only leftover chain.

"MOTORCYCLE #1"
BY BRUCE GRAY

The only full-sized motorcycle sculpture ever produced by the renowned sculptor Bruce Gray, using industrial parts and an actual BMW R75 engine.

LIFE-SIZE BRONZE SCULPTURES BY SERGE BUENO

Commissioned by Bobby Haas and the Haas Moto Museum, Serge Bueno disassembled five of both actual motorcycles and models to create molds into which molten bronze was poured and then welded together. The final result was a series of five life-size, 1,000-pound bronze sculptures, including a c.1911 Flying Merkel Factory Racer, a 1915 Harley-Davidson 11-K Board Track Racer, a 1915 Cyclone Board Track Racer, a 1917 Henderson Four Racer, and a c.1918 Indian Board Track Racer.

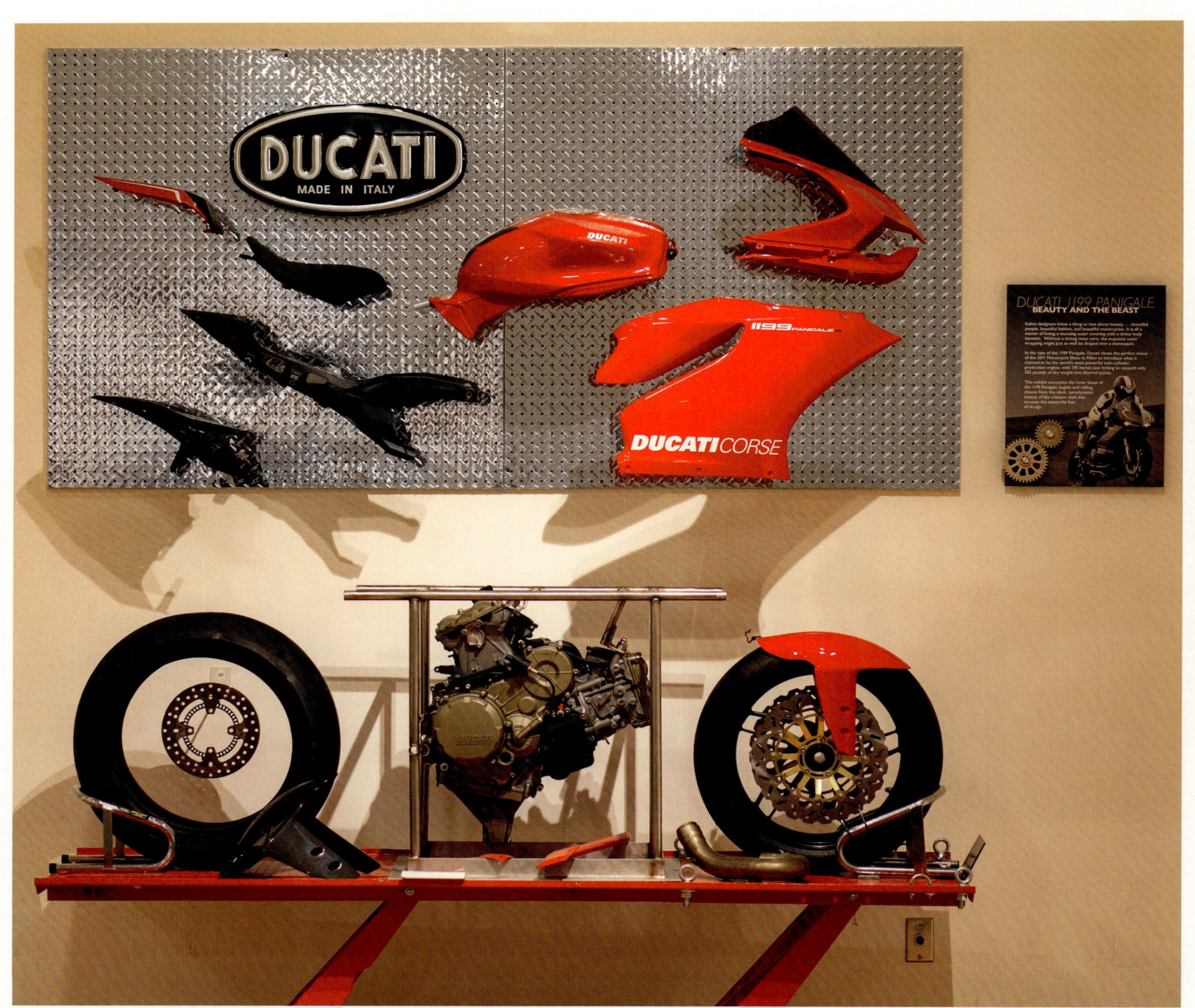

"EXPLODING DUCATI"

Ducati chose the perfect venue of the 2011 Motorcycle Show in Milan to introduce what it claimed to be the world's most powerful twin-cylinder production engine in the 1199 Panigale, with 195 horses just itching to catapult only 362 pounds of dry weight into blurred action. This exhibit uncouples the inner beast of the 1199 Panigale engine and rolling chassis from the sleek, aerodynamic beauty of the crimson shell that encases this masterful feat of design.

"RONDINE" BY MEDAZA CYCLES

Guzzi Falcone 1/1... This sculpture represents the Rondine by Medaza Cycles in the Custom Shop, the custom cycle that captured the 2013 AMD World Championship in Essen, Germany.

Favorit

ABOUT BOBBY HAAS

For Robert B. Haas (known only as "Bobby"), the kindest compliment of all was to be called a "Renaissance man"—a person who pushes to expand the limits of success in one enterprise after another, usually with no pedigree in the field and no rightful claim to succeed. He believed a Renaissance man must forever reinvent himself after he has achieved a measure of success in one pursuit, starting over each time with none of the privileges or the accoutrements of the previous success to provide a tailwind for the next venture. Like a biker accelerating from a standing start, a Renaissance man must overcome the headwinds first encountered and then achieve a cruising speed with practiced skills until the next destination is reached.

After graduating with honors from Yale in 1969 and from Harvard Law School in 1972, Bobby found early fortunes through the leveraged buyout business in Dallas, Texas, in the 1980s, including the notable sale of a soft-drink empire that included brands such as Dr Pepper, A&W, and 7UP. He often described his business success as giving him a "bulletproof vest" to pursue his passions and art.

In his late forties, Bobby made his first entry into the art world by taking up photography. Soon he became one of the world's most celebrated aerial photographers and the creator of three of the most widely distributed works in the storied history of *National Geographic.*

The artistic pursuit didn't stop there. Just a few years after kick-starting his first motorcycle at the age of sixty-four (an all-black 2011 Ural Retro), Bobby set his sights on assembling a collection that would be exhibited in a museum like no other—in his words, "pound-for-pound, the most stunning exhibition of motorcycle artistry anywhere." With a collection that now numbers more than 230 cycles and sculptures curated in just over five years, it is clear that the passion became a dream, and the dream became a mission.

Every edifice to the arts must have a guiding spirit, a catalyst for its creation. But rarely has one person exerted such a pervasive influence on a museum by taking on such a multiplicity of roles—founder and driving creative spirit; sole benefactor of the entire enterprise; curator of the collection; designer of custom cycles, sculptures, and furnishings; and supervisor of construction. Bobby has his fingerprints in every nook and cranny of the Haas Moto Museum, and his infectious passion for the museum has inspired every member of the creative and operational team.

As Bobby always said, "If you only live once, you might as well live a bunch of lives."

ACKNOWLEDGMENTS

Thank you to those that have participated in this book project. Appreciation runs deep for those who offered guidance, enthusiasm, support, patience, determination, creativity, and time.

The Haas Team: Brent, Daniel, Dylan, Nancy and Sparky

Dylan Guest

Shaik Ridzwan

Grant Schwingle

Karyn Gerhard

The NT Team

Dupree Miller

PHOTO CREDITS

All photography of motorcycles and museum staff are courtesy of **Grant Schwingle**, with the exception of those listed below.
(Key: R=right; L=left; C=center; T=top; B=bottom)

Brandon LaJoie: 252; **Brent Graves**: 8, 10, 229T, 229BL, 229BR, 240, 241TL, 241TR, 241CL, 241CR, 241BL, 241BR, 242, 243, 244T, 244B, 245, 246, 247, 248; **Shaik Ridzwan**: 11T.

DEDICATION

This book is dedicated to Team Haas, the small but mighty museum family that Bobby built. Because of the devotion, commitment, and love of this loyal group, Bobby's vision, passion, and legacy will endure for generations to come.

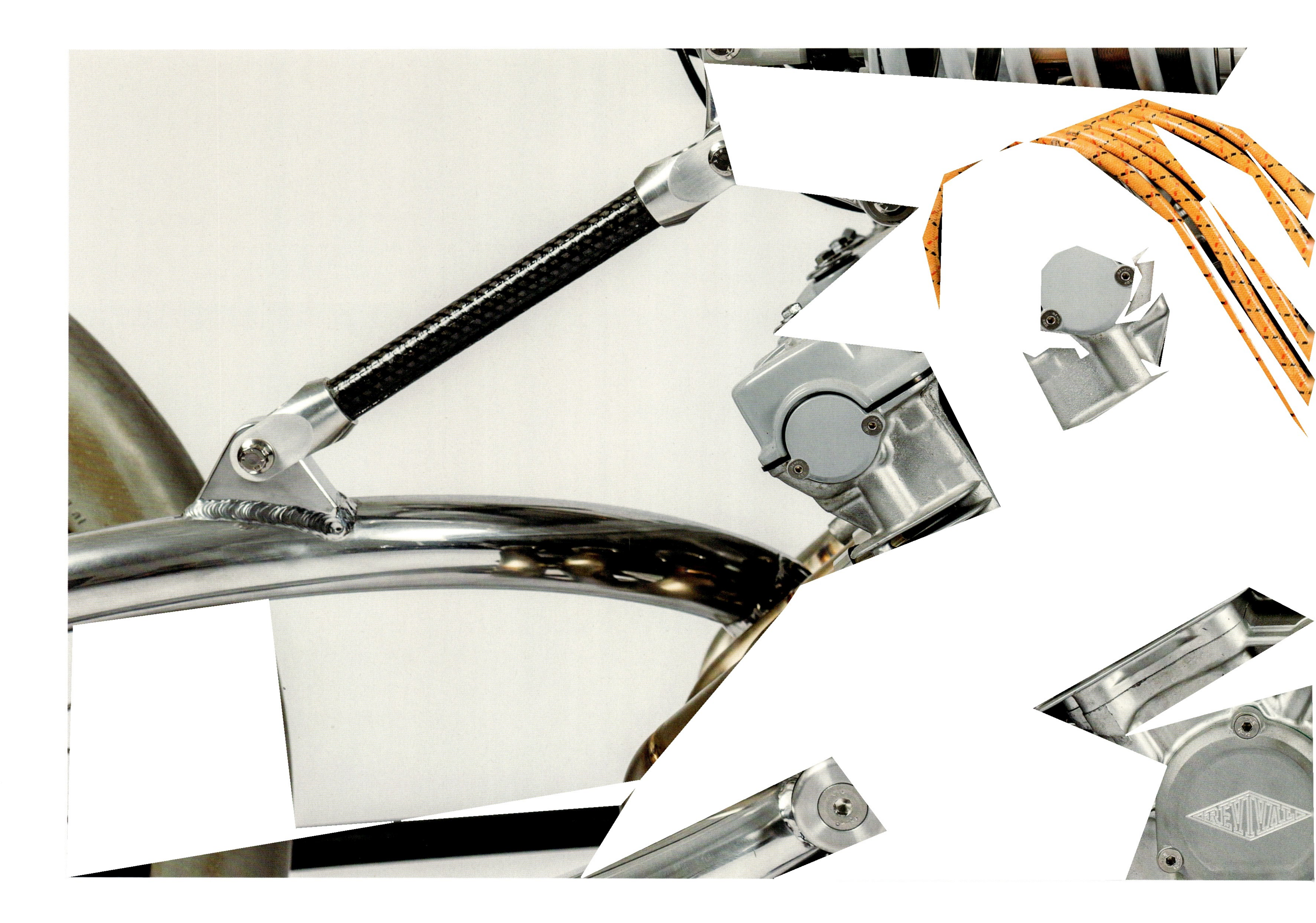
REVIVAL

weldon**owen**

Insight Editions
PO Box 3088
San Rafael, CA 94912
www.insighteditions.com

Find us on Facebook: www.facebook.com/InsightEditions

Follow us on Twitter: @insighteditions

CEO Raoul Goff
Publisher Roger Shaw
Editorial Director Katie Killebrew
Senior Editor Karyn Gerhard
VP Creative Chrissy Kwasnik
Art Director Allister Fein
VP Manufacturing Alix Nicholaeff
Sr Production Manager Joshua Smith
Sr Production Manager, Subsidiary Rights Lina s Palma-Temena

Design by Roger Gorman, Reiner Design Consultants Inc.

Weldon Owen would like to thank Jon Ellis and Karen Levy for their extraordinary work on this book.

ISBN: 979-8-88674-047-9

Manufactured in China by Insight Editions
10 9 8 7 6 5 4 3 2 1